T0140446

Studies in Fuzziness and Soft Computing

Volume 324

Series editor

Janusz Kacprzyk, Polish Academy of Sciences, Warsaw, Poland
e-mail: kacprzyk@ibspan.waw.pl

About this Series

The series "Studies in Fuzziness and Soft Computing" contains publications on various topics in the area of soft computing, which include fuzzy sets, rough sets, neural networks, evolutionary computation, probabilistic and evidential reasoning, multi-valued logic, and related fields. The publications within "Studies in Fuzziness and Soft Computing" are primarily monographs and edited volumes. They cover significant recent developments in the field, both of a foundational and applicable character. An important feature of the series is its short publication time and world-wide distribution. This permits a rapid and broad dissemination of research results.

More information about this series at http://www.springer.com/series/2941

Anjan Mukherjee

Generalized Rough Sets

Hybrid Structure and Applications

 Springer

Anjan Mukherjee
Department of Mathematics
Tripura University
Agartala, Tripura
India

ISSN 1434-9922   ISSN 1860-0808   (electronic)
Studies in Fuzziness and Soft Computing
ISBN 978-81-322-3441-8   ISBN 978-81-322-2458-7   (eBook)
DOI 10.1007/978-81-322-2458-7

Preface

The book is designed for researchers and students working in the field of fuzzy set, rough set, soft set, and their hybrid models. In our real-life problems, there are situations with the uncertain data that may not be successfully modelled by the classical mathematics. There are some mathematical tools for dealing with uncertainties: fuzzy set theory, rough set theory, and soft set theory. The book is written in order to accumulate all the contents of generalised fuzzy theory and all hybrid structures of fuzzy set, rough set, and soft set, so that the researchers get all the information at one place. The primary goal of this book is to help bridge the gap to provide a textbook on the hybrid structures in fuzzy mathematics and their applications in social science.

The concept of 'fuzzy set theory' was first introduced by Lotfi A. Zadeh in 1965 (*Information and Control*, vol. 8, pp. 338–353) and thereafter by C.L. Chang (in Fuzzy topological spaces, *J. Math. Anal. Appl.*, vol. 24, pp. 182–190), paved the way of subsequent development of numerous fuzzy topological concepts. In 1983, Atanassov introduced the concept of 'intuitionistic fuzzy set' as a generalisation of the notion of a fuzzy set. Intuitionistic fuzzy sets give both a degree of membership and a degree of non-membership, which are independent of each other. The only requirement is that the sum of these two degrees is not greater than 1. Using intuitionistic fuzzy sets, not only vagueness but also uncertainty is modelled. The concept of 'rough set theory', which was first introduced by Z. Pawlak in 1981/1982, deals with the approximation of sets that are difficult to describe with the available information. Rough set introduced by Z. Pawlak is expressed by a boundary region of a set. It is also an approach to vagueness. Thus, fuzzy sets and rough sets are two different approaches to vagueness or impreciseness of the real-life problems. The 'soft set theory', which was introduced by Molodtsov in 1999, takes care of the problem that involves such vagueness. In 2001, Maji et al. introduced the idea of intuitionistic fuzzy soft set theory and established some results on them. Theories of fuzzy sets and rough sets are powerful mathematical tools for modelling various types of uncertainty. Molodtsov [2] initiated a novel concept called soft sets, a new mathematical tool for dealing with uncertainties. It has been found that fuzzy set, rough set, and soft set are closely related concepts [1]. Research works on soft sets are very active and progressing rapidly in these years.

The book introduces the concept and basic properties of 'generalised interval-valued intuitionistic fuzzy soft sets'. It also discusses the application of generalised interval-valued intuitionistic fuzzy soft sets in decision making with respect to interval of degree of preference. The book also introduces the concept of 'soft rough intuitionistic fuzzy sets' and 'interval-valued intuitionistic fuzzy soft rough sets'. The concept of interval-valued intuitionistic fuzzy soft rough set-based multi-criteria group decision-making scheme is presented, which refines the primary evaluation of the whole expert group and enables us to select the optimal object in a most reliable manner. The proposed scheme is to be illustrated by an example regarding the candidate selection problem.

The book also introduces the concept of 'interval-valued intuitionistic fuzzy soft topological space (IVIFS topological space)' together with intuitionistic fuzzy soft open sets (IVIFS open sets) and intuitionistic fuzzy soft closed sets (IVIFS closed sets). Then, we define neighbourhood of an IVIFS set, interior IVIFS set, interior of an IVIFS set, exterior IVIFS set, exterior of an IVIFS set, closure of an IVIFS set, IVIF soft basis, and IVIFS subspace. Some examples and theorems regarding these concepts are presented. The concepts of 'interval-valued intuitionistic fuzzy soft multi-sets' and that of 'interval-valued intuitionistic neutrosophic soft sets' are introduced and their applications are studied.

The book is organised in ten chapters. In Chap. 1, the basic concepts of fuzzy set, rough set, soft set, fuzzy topology, and intuitionistic fuzzy topology are given for ready reference. Also relation among fuzzy set, rough set, and soft set are shown with suitable examples. Here, soft multi-set and the concept of topological structure formed by soft multi-sets are also discussed. One of the areas in which fuzzy sets, rough sets, and soft sets have been applied most extensively is in modelling for managerial decision making. Thus, this area has been selected by us for consideration. In Chap. 2, the concept of generalised interval-valued intuitionistic fuzzy soft sets together with their basic properties is introduced. Besides, an application of generalised interval-valued intuitionistic fuzzy soft sets in decision making is also presented. In Chap. 3, soft rough intuitionistic fuzzy sets (SRIF sets) are also discussed. Finally, an example that demonstrated that this method can be successfully work is also provided here. It can be applied to problems of many fields that contain uncertainty. The aim of Chap. 4 is to introduce a new concept—interval-valued intuitionistic fuzzy soft rough sets. Also, interval-valued intuitionistic fuzzy soft rough set-based multi-criteria group decision-making scheme is presented. In Chap. 5, the concept of interval-valued intuitionistic fuzzy soft topology is introduced and their basic properties are studied. Chapter 6 introduces the concept of interval-valued intuitionistic fuzzy soft multi-sets and study of its properties and operations. The concept of interval-valued intuitionistic fuzzy soft multi-set relations (IVIFSMS relations for short) is also proposed. Besides, the basic properties of the IVIFSMS relations and various types of IVIFSMS relations are presented in this chapter. Chapter 7 introduces the concept of interval-valued neutrosophic set (IVNS), interval-valued neutrosophic soft set (IVNSS), and interval-valued neutrosophic soft set (IVNSS) relations. Besides, the basic properties of IVNSS and IVNSS relations and various types of IVNSS relations are studied and also discussed in this chapter. In Chap. 8, the concept of fuzzy soft multi-topology is introduced and their basic properties are studied. The notion of fuzzy soft multi-points, fuzzy soft multi-open sets, fuzzy soft multi-closed sets, fuzzy soft multi-basis, fuzzy soft multi-sub basis, neighbourhoods and neighbourhood systems, and interior and closure of a fuzzy soft multi-set are introduced and their basic properties are investigated. It is shown that a fuzzy soft multi-topological space gives a parameterised family of fuzzy topological spaces. Here, the concepts of fuzzy soft multi-topological subspaces and some basic properties of these concepts are also explored. Lastly, the concept of fuzzy soft multi-compactness and fuzzy soft multi-separation axioms is introduced. The aim of Chap. 9 was to introduce the concept of soft interval-valued intuitionistic fuzzy rough sets. Also, some properties based on soft interval-valued intuitionistic fuzzy rough sets are presented here. Finally, a soft interval-valued intuitionistic fuzzy rough set-based multi-criteria group decision-making scheme is presented. The proposed scheme is illustrated by an example regarding the car selection problem. The aim of the Chap. 10 is to construct IF parameterised intuitionistic fuzzy soft set (IFPIFS set) decision-making problem and to solve the problem with IF parameterised intuitionistic fuzzy soft set theories.

In our real-life problems, there are situations with the uncertain data that may not be successfully modelled by the classical mathematics. There are some mathematical tools for dealing with uncertainties; they are fuzzy set theory introduced by Prof. Zadeh [4], rough set theory introduced by Pawlak [3], and soft set theory initiated by Molodtsov [2] that are related to our work.

I gratefully acknowledge the support provided by Springer for publishing this book.

Prof. Anjan Mukherjee

References

1. Aktas, H., Cagman, N.: Soft sets and soft groups. Inf. Sci. **177**(13), 2726–2735 (2007)
2. Molodtsov, D.A.: Soft set theory—first results. Comput. Math. Appl. **37**(4–5), 19–31 (1999)
3. Pawlak, Z.: Rough sets. Int. J. Inf. Comput. Sci. **11**, 341–356 (1982)
4. Zadeh, L.A.: Fuzzy sets. Inf. Control. **8**, 338–353 (1965)

Contents

About the Author

Prof. Anjan Mukherjee is Pro Vice-Chancellor of Tripura University. He has completed his B.Sc and M.Sc in mathematics from University of Calcutta and obtained his Ph.D. from Tripura University. Dr. Mukherjee has 26 years of vast experience in research and teaching. He has published more than 150 research papers in different national and international journals and conference proceedings and has delivered several invited talks. Dr. Mukherjee is in the Editorial Board of *Universal Journal of Computational Mathematics* and also associated with Fuzzy and Rough Sets Association. He has presented his work at University of Texas (USA), City College of New York (USA), 5th Asian Mathematical Conference (Malaysia), and many other universities.

Chapter 1
Introduction to Fuzzy Sets, Rough Sets, and Soft Sets

Abstract In our real-life problems, there are situations with the uncertain data that may not be successfully modelled by the classical mathematics. There are some mathematical tools for dealing with uncertainties—they are fuzzy set theory introduced by Zadeh [10], rough set theory introduced by Pawlak [7], and soft set theory initiated by Molodtsov [5]. In this chapter, we recall some basic notions relevant to our Chaps. 2–10, such as fuzzy sets, intuitionistic fuzzy sets, interval fuzzy sets, soft set, fuzzy soft sets, rough sets, fuzzy rough sets, fuzzy rough soft set, and others.

Keywords Fuzzy set · Rough set · Soft set · Soft multi-set · Interval valued fuzzy set · Intuitionistic fuzzy set · Fuzzy rough set

In our real-life problems, there are situations with the uncertain data that may not be successfully modelled by the classical mathematics. There are some mathematical tools for dealing with uncertainties—they are fuzzy set theory introduced by Zadeh [10], rough set theory introduced by Pawlak [7], and soft set theory initiated by Molodtsov [5]. A fuzzy set allows a membership value other than 0 and 1. A rough set uses there membership functions, a reference set and its lower and upper approximations in an approximation space. There are extensive studies on the relationships between rough sets and fuzzy sets. Many proposals have been made for the combination of rough set and fuzzy set. The 'soft set theory', which was introduced by Molodtsov in 1999 [5], takes care of the problem that involves such vagueness.

1.1 Sets and Subsets

In the classical or ordinary set theory, a set is a 'well-defined' collection of objects. By 'well defined', we mean that there is a given rule by means of which it is possible to know whether a particular object is contained in the collection or not.

Let X be an ordinary set and A a subset of it. We write $x \in A$ if an element x of X is a member of A and $x \notin A$ if x of X is not a member of A. Membership in a subset A of X is based on two-valued logic and can be restated in terms of characteristic function (or membership function) μ_A from X to $\{0, 1\}$, i.e.

$$\mu_A(x) = 1, \quad \text{if } x \in A$$
$$= 0, \quad \text{if } x \notin A$$

Let us consider a finite set $X = \{a, b, c, d, e, f\}$ and a subset $A = \{a, b, d, f\}$.
A can be represented by the set of pairs, i.e.

$$A = \{(x, \mu_A(x)) : x \in X\}$$
$$= \{(a, 1), (b, 1), (c, 0), (d, 1), (e, 0), (f, 1)\}$$

where μ_A is the known characteristic function and $\mu_A \colon X \to \{0, 1\}$.

Thus, the subset A of a set X can be characterised by a characteristic function which associates with each x its grade of membership $\mu_A(x)$, i.e.

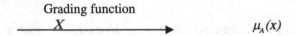

$$\xrightarrow[\quad\quad X \quad\quad]{\text{Grading function}} \quad\quad \mu_A(x)$$

1.2 Fuzzy Subsets

Fuzzy set is a generalisation of the classical set or crisp set.

Examples of fuzzy set	Examples of crisp set
1. Tall men	1. Men with height ≥5 ft.
2. Good boys	2. Boys passed in 1st class
3. Red flowers	3. All roses
4. Young	4. Person with age ≤10 years
5. Expensive cars	5. Cars with price more than 5 lacs
6. Adults	6. Persons with age ≥18 years
7. Sunny days	7. Days between 15 April and 31 May
8. Olds	8. Persons with age ≥60
9. All points near 5	9. All points in [4.7, 5.3]

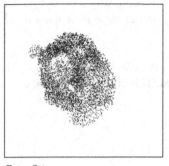 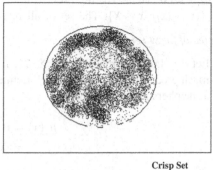

Fuzzy Set   **Crisp Set**

Boundary is not sharp   **Boundary is sharp**

Fuzzy sets can be applied in the following fields: engineering, psychology, medicine, artificial intelligence, ecology, decision-making theory, pattern recognition, sociology, meteorology, computer science, manufacturing, and others. It occurs in all the areas in which human judgement evaluation and decision are important.

Let us consider the following two subfamilies of a set X of students

(i) $B = \{$a collection of all students of Tripura University$\}$
(ii) $C = \{$a collection of all intelligence students of Tripura University$\}$

In example (i), one can be sure whether a particular member belongs to the collection or not. Thus, it is a collection of well-defined and distinct objects, and hence, it is a set. The membership in the subset B of X is defined by

$$\mu_B(x) = 1, \quad \text{if } x \in A$$
$$\qquad\ = 0, \quad \text{if } x \notin A$$

In example (ii), one cannot be sure whether a particular member belongs to the collection or not. The subfamily of the above kind is not precise, and the simplest way to describe the above collection mathematically is to characterise the degree of belongingness by a number from the closed interval [0, 1].

Definition 1.1 Let X be an ordinary set. A fuzzy subset α in X is the collection of ordered pairs $(x, \mu_\alpha(x))$ with $x \in X$ and a membership function $\mu_\alpha: X \to [0, 1]$. The value $\mu_\alpha(x)$ of x denotes the degree to which an element x may be a member of α. Thus, a fuzzy subset α of X is denoted by $\alpha = \{(x, \mu_\alpha(x)): x \in X\}$ where $\mu_\alpha(x) = 1$ indicates strictly the containment of the element x in α (full membership) and $\mu_\alpha(x) = 0$ denotes that x does not belong to α (non-membership). Thus, an ordinary set is a special case of fuzzy set with a membership function which is reduced to a characteristic function. Because of these generalities, the fuzzy set theory has a wider scope of applicability than the ordinary set theory in solving real problem.

A fuzzy set α can also be represented in the following way $\alpha = \{x/\mu_\alpha(x), \forall x \in X\}$ or $\alpha = \{(x, \mu_\alpha(x)): x \in X\}$. The set of all fuzzy subset on X is denoted by I^X.

Examples of fuzzy sets 1.2

(i) Let $\alpha = \{x/x \text{ is a real number} \gg 1\}$, i.e. the class of all numbers x that are much greater than 1. It is not well defined. This collection may be defined by a membership function where

$$\mu_\alpha(x) = 0$$
$$= (x - 1)/x$$

Thus, the fuzzy subset

$$\alpha = \{(1,0), (2,0.5), (4,0.75), \ldots, \ldots, \ldots (10,0.9), \ldots \ldots (100,0.99), \ldots \ldots\}$$

(ii) Unemployment is a fuzzy concept; we know that there are people with part-time job who would like to work full time. This collection may be defined in the following set of unemployed people.

$$\{(0,1), \ldots, (4,0.9), \ldots (20,0.5), \ldots (40,0)\}$$

where the number in the 1st coordinate indicates the hours worked per week and the number in the 2nd coordinate indicates the degree to which the person is unemployed, in which 1 means unemployed, 0 means employed and 0.5 means 50 % unemployed.

1.3 Basic Operations on Fuzzy Subsets

Let α and β be two fuzzy subsets of X with the membership function μ_α and μ_β, respectively. Then, for all $x \in X$, we have

(i) α is equal to β, i.e. $\alpha = \beta$ iff $\mu_\alpha(x) = \mu_\beta(x)$
(ii) α is a subset of β, i.e. $\alpha \leq \beta$ iff $\mu_\alpha(x) \leq \mu_\beta(x)$
(iii) Union of α and β, i.e. $\alpha \vee \beta$ iff $\mu_{\alpha\vee\beta}(x) = \max\{\mu_\alpha(x), \mu_\beta(x)\}$
(iv) Intersection of α and β, i.e. $\alpha \wedge \beta$ iff $\mu_{\alpha\wedge\beta}(x) = \min\{\mu_\alpha(x), \mu_\beta(x)\}$
(v)   Complement of β, i.e. $\beta^c = 1 - \beta$ iff $\mu_\beta^c(x) = 1 - \mu_\beta(x)$

It can be seen that commutative laws, associative laws, distributive laws, De Morgan's laws, and idempotent laws are true for fuzzy subsets.

For a family of fuzzy subsets $\{\alpha_j: j \in \Lambda\}$

$$\gamma = \bigvee_{j\in\Lambda} \alpha_j \Leftrightarrow \mu_\alpha(x) = \sup_{j\in\Lambda}\{\mu_{\alpha_j}(x): x \in X\}$$

and

$$\eta = \bigwedge_{j \in \Lambda} \alpha_j \Leftrightarrow \mu_\alpha(x) = \inf_{j \in \Lambda}\left\{\mu_{\alpha_j}(x): x \in X\right\}$$

Example 1.3 Let $X = \{x_1, x_2, x_3, x_4, x_5, x_6\}$ and the two fuzzy subsets of X be

$$\alpha = \{(x_1, 0.6), (x_2, 0.4), (x_3, 0.3), (x_4, 0.8), (x_5, 0.5), (x_6, 1)\},$$
$$\beta = \{(x_1, 0.8), (x_2, 0.3), (x_3, 1), (x_4, 1), (x_5, 0.4), (x_6, 0.9)\}$$

Then,

$$\beta^c = \{(x_1, 0.2), (x_2, 0.7), (x_3, 0), (x_4, 0), (x_5, 0.6), (x_6, 0.1)\}$$

Now,

$$\alpha \vee \beta = \{(x_1, 0.8), (x_2, 0.4), (x_3, 1), (x_4, 1), (x_5, 0.5), (x_6, 1)\}$$
$$\alpha \wedge \beta = \{(x_1, 0.6), (x_2, 0.3), (x_3, 0.3), (x_4, 0.8), (x_5, 0.4), (x_6, 0.9)\}$$

If $A = \{x_6\}$ and $B = \{x_3, x_4\}$ are the two ordinary subsets of X, then $B^c = \{x_1, x_2, x_5, x_6\}$ and $B \cup B^c = X$ and $B \cap B^c = \phi$. But in fuzzy set theory,

$$\beta \vee \beta^c = \{(x_1, 0.8), (x_2, 0.7), (x_3, 1), (x_4, 1), (x_5, 0.6), (x_6, 0.9)\} \neq X$$
$$= \{(x_1, 1), (x_2, 1), (x_3, 1), (x_4, 1), (x_5, 1), (x_6, 1)\}$$

And

$$\beta \wedge \beta^c = \{(x_1, 0.2), (x_2, 0.3), (x_3, 0), (x_4, 0), (x_5, 0.4), (x_6, 0.1)\} \neq \phi$$
$$= \{(x_1, 0), (x_2, 0), (x_3, 0), (x_4, 0), (x_5, 0), (x_6, 0)\}$$

After the introduction of concept of fuzzy sets by Zadeh, several researches were conducted on the generalisation of the notion of the fuzzy set. The idea of 'intuitionistic fuzzy set' was first published by Atanassov (1983).

Definition 1.4 Let a set E be fixed. An **intuitionistic fuzzy set or IFS** 'A' in E is an object having the form $A = \{(x, \mu_A(x), \nu_A(x): x \in E)\}$ where the functions μ_A: $E \rightarrow I = [0, 1]$ & ν_A: $E \rightarrow I = [0, 1]$ define the degree of membership and non-membership, respectively, of the element $x \in E$ to the set A & for every $x \in E$, $0 \leq \mu_A(x) + \nu_A(x) \leq 1$. The rest part $\pi_A(x) = 1 - \mu_A(x) - \nu_A(x)$ is called the indeterministic part of x and $0 \leq \pi_A(x) \leq 1$.

Definition 1.5 Let X be a non-empty set and the IF sets A and B be in the form $A = \{(x, \mu_A(x), \nu_A(x): x \in X)\}$, $B = \{(x, \mu_B(x), \nu_B(x): x \in X)\}$.

Then

(a) $A \subseteq B$ if and only if $\mu_A(x) \leq \mu_B(x)$ and $v_A(x) \geq v_B(x)$ for all $x \in X$.
(b) $A = B$ if and only if $A \subseteq B$ and $B \subseteq A$.
(c) $A^c = \{(x, v_A(x), \mu_A(x): x \in X)\}$.
(d) $A \cap B = \{(x, \mu_A(x) \cap \mu_B(x), v_A(x) \cup v_B(x): x \in X)\}$.
(e) $A \cup B = \{(x, \mu_A(x) \cup \mu_B(x), v_A(x) \cap v_B(x): x \in X)\}$.

Definition 1.6 An *interval-valued fuzzy set* A over a universal set U is defined as the object of the form $A = \{(x, \mu_A(x): x \in U)\}$, where $\mu_A(x): U \to \text{Int}([0, 1])$ is a function, where $\text{Int}([0, 1])$ denotes the set of all closed subintervals of $[0, 1]$.

Definition 1.7 An *interval-valued intuitionistic fuzzy set* A over a universal set U is defined as the object of the form $A = \{\langle x, \mu_A(x), \gamma_A(x)\rangle : x \in U)\}$ where $\mu_A(x): U \to \text{Int}([0, 1])$ and $\gamma_A(x): U \to \text{Int}([0, 1])$ are functions such that the condition $\forall x \in U$, $\sup \mu_A(x) + \sup \gamma_A(x) \leq 1$ is satisfied.

The class of all interval-valued intuitionistic fuzzy soft sets on U is denoted by IVIFS^U. For an arbitrary set $A \subseteq [0, 1]$, we use $\underline{A} = \inf A$ and $\overline{A} = \sup A$.

Definition 1.8 Let R be an equivalence relation on the universal set U. Then, the pair (U, R) is called a Pawlak approximation space. An equivalence class of R containing x will be denoted by $[x]_R$. Now, for $X \subseteq U$, the lower and upper approximations of X with respect to (U, R) are denoted by, respectively, R_*X & R^*X and are defined by

$$R_*X = \{x \in U: [x]_R \subseteq X\} \quad \text{or} \quad \bigcup_{x \in U}\{[x]_R: [x]_R \subseteq X\},$$

$$R^*X = \{x \in U: [x]_R \cap X \neq \phi\} \quad \text{or} \quad \bigcup_{x \in U}\{[x]_R: [x]_R \cap X \neq \phi\},$$

Now, if $R_*X = R^*X$, then X is called definable; otherwise, X is called a **rough set**.

Example 1.9 Let $A = (U, R)$ be an approximate space where $U = \{0, 1, 2, 3, 4, 5, 6, 7, 8, 9, 10\}$ and the relation R on U be definable aRb iff $a \equiv b \pmod 5$ for all a, $b \in U$. Let us consider a subset $X = \{1, 2, 6, 7, 8, 9\}$ of U.

Then, the rough set of X is $A(x) = (\overline{A}(x), \underline{A}(x))$ where $\underline{A}(x) = \{1, 2, 6, 7\}$ and $\overline{A}(x) = \{1, 2, 3, 4, 6, 7, 8, 9\}$.

Here, the equivalence classes are

$[0]_R = [5]_R = [10]_R = \{0, 5, 10\}$
$[1]_R = [6]_R = \{1, 6\}$
$[2]_R = [7]_R = \{2, 7\}$
$[3]_R = [8]_R = \{3, 8\}$
$[4]_R = [9]_R = \{4, 9\}$

Thus, $\underline{A}(x) = \{x \in U: [x]_R \subset X\} = \{1, 2, 6, 7\}$.

$$\overline{A}(x) = \{x \in U: [x]_R \cap X \neq \phi\} = \{1,2,3,4,6,7,8,9\}$$

Let us now define the cardinality of $[0]_R, [1]_R, [2]_R, [3]_R, [4]_R$

i.e. cardinality of $[0]_R$ is $\square$ $[0]_R = 3 = \square[5]_R = \square[10]_R$
cardinality of $[1]_R$ is $\square$ $[1]_R = 2 = \square[6]_R$
cardinality of $[2]_R$ is $\square$ $[2]_R = 2 = \square[5]_R$
cardinality of $[3]_R$ is $\square$ $[3]_R = 2 = \square[8]_R$
cardinality of $[4]_R$ is $\square$ $[4]_R = 2 = \square[9]_R$

Now, for any element, $u \in U$ degree of rough belongingness of u in X is $\square$ $\{[u]_R \cap X\}/[u]_R$ and is denoted by $\mu^R_{F_X}(u)$

So $\mu^R_{F_X}(0) = \square \{[0]_R \cap X\}/[0]_R = \square\phi/3 = 0$
$\mu^R_{F_X}(1) = \square \{[1]_R \cap X\}/[1]_R = \square\{1, 6\}/2 = 2/2 = 1$
$\mu^R_{F_X}(2) = \square \{[2]_R \cap X\}/[2]_R = \square\{2,7\}/2 = 2/2 = 1$
$\mu^R_{F_X}(3) = 0.5, \mu^R_{F_X}(4) = 0.5, \mu^R_{F_X}(5) = 0, \mu^R_{F_X}(6) = 1, \mu^R_{F_X}(7) = 1, \mu^R_{F_X}(8) = 0.5,$
$\mu^R_{F_X}(9) = 0.5, \mu^R_{F_X}(10) = 0$

Hence, $F^R_X = \{(0,0),(1,1),(2,1),(3,0.5),(4,0.5),(5,0),(6,1),(7,1),(8,0.5),(9,0.5),$ $(10,0)\}$ induces a fuzzy set F^R_X of U.

Thus, $F^R_X = \left\{\left(u, \mu^R_{F_X}(u)\right): u \in U\right\}$ is the fuzzy set.

Remark 1.10 It is interesting to compare definitions of classical sets, fuzzy sets, and rough sets. Classical set is a primitive notion and is defined intuitively or axiomatically. Fuzzy sets are defined by employing the fuzzy membership function, which involves advanced mathematical structures, numbers, and functions. Rough sets are defined by approximations. Thus, this definition also requires advanced mathematical concepts.

Approximations have the following properties:

1. $R_*(X) \subseteq X \subseteq R^*(X)$
2. $R_*(\varnothing) = R^*(\varnothing) = \varnothing; R_*(U) = R^*(U) = U$
3. $R^*(X \cup Y) = R^*(X) \cup R^*(Y)$
4. $R_*(X \cap Y) = R_*(X) \cap R_*(Y)$
5. $R_*(X \cup Y) \supseteq R_*(X) \cup R_*(Y)$
6. $R^*(X \cap Y) \subseteq R^*(X) \cap R^*(Y)$
7. $X \subseteq Y \rightarrow R_*(X) \subseteq R_*(Y) \& R^*(X) \subseteq R^*(Y)$
8. $R_*(-X) = -R^*(X)$
9. $R^*(-X) = -R_*(X)$
10. $R_* R_*(X) = R^* R_*(X) = R_*(X)$
11. $R^* R^*(X) = R_* R^*(X) = R^*(X)$

It is easily seen that approximations are in fact interior and closure operations in a topology generated by data. Thus, fuzzy set theory and rough set theory require completely different mathematical setting.

Rough sets can be also defined employing, instead of approximation, rough membership function.

Definition 1.11 Let U be an initial universe and E be a set of parameters. Let $P(U)$ denote the power set of U and $A \subseteq E$. Then, the pair (F, A) is called a *soft set* over U, where F is a mapping given by $F: A \rightarrow P(U)$.

For any $\varepsilon \in A$, $F(\varepsilon) \subseteq U$ may be considered as the set of ε-approximate elements of the soft set (F, A).

Definition 1.12 Let U be an initial universe and E be a set of parameters. Let I^U be the set of all fuzzy subsets of U and $A \subseteq E$. Then, the pair (F, A) is called a *fuzzy soft set* over U, where F is a mapping given by $F: A \rightarrow I^U$.

For any $\varepsilon \in A$, $F(\varepsilon)$ is a fuzzy subset of U. Let us denote the membership degree that object x holds parameter ε by $\mu_{F(\varepsilon)}(x)$, where $x \in U$ and $\varepsilon \in A$. Then, $F(\varepsilon)$ can be written as a fuzzy set such that $F(\varepsilon) = \{(x, \mu_{F(\varepsilon)}(x)): x \in U\}$.

Definition 1.13 Let U be an initial universe and E be a set of parameters. Let I^U be the set of all fuzzy subsets of U and $A \subseteq E$. Now, $F: A \rightarrow I^U$, and α be a fuzzy subset of A, i.e. $\alpha: A \rightarrow I = [0, 1]$. Let F_α be a mapping $F_\alpha: A \rightarrow I^U \times I$ defined as follows: $F_\alpha(\varepsilon) = (F(\varepsilon), \alpha(\varepsilon))$, and then, F_α is called **generalised fuzzy soft set** over the soft universe (U, A).

For any $\varepsilon \in A$, $F(\varepsilon)$ is a fuzzy subset of U. Let us denote the membership degree that object x holds parameter ε by $\mu_{F(\varepsilon)}(x)$, where $x \in U$ and $\varepsilon \in A$. Then, $F(\varepsilon)$ can be written as a fuzzy set such that $F(\varepsilon) = \{(x, \mu_{F(\varepsilon)}(x)): x \in U\}$.

Definition 1.14 Let U be an initial universe and E be a set of parameters. Let IF^U be the set of all intuitionistic fuzzy subsets of U and $A \subseteq E$. Then, the pair (F, A) is called an *intuitionistic fuzzy soft set* over U, where F is a mapping given by $F: A \rightarrow IF^U$.

For any $\varepsilon \in A$, $F(\varepsilon)$ is an intuitionistic fuzzy subset of U. Let us denote $\mu_{F(\varepsilon)}(x)$ and $\gamma_{F(\varepsilon)}(x)$ by the membership degree and non-membership degree, respectively, that object x holds parameter ε, where $x \in U$ and $\varepsilon \in A$. Then, $F(\varepsilon)$ can be written as an intuitionistic fuzzy set such that $F(\varepsilon) = \{(x, \mu_{F(\varepsilon)}(x), \gamma_{F(\varepsilon)}(x)): x \in U\}$.

Definition 1.15 Let U be an initial universe and E be a set of parameters. Let IF^U be the set of all intuitionistic fuzzy subsets of U and $A \subseteq E$. Let F be a mapping given by $F: A \rightarrow \mathrm{IF}^U$ and α be a mapping given by $\alpha: A \rightarrow [0, 1]$. Let F_α be a mapping given by

$F_\alpha: A \rightarrow \mathrm{IF}^U \times [0, 1]$ and defined by

$$F_\alpha(e) = (F(e), \alpha(e))$$
$$= \left(\left\langle x, \mu_{F(e)}(x), \gamma_{F(e)}(x) \right\rangle, \alpha(e) \right),$$

where $e \in A$ and $x \in U$.

Then, the pair (F_α, A) is called a **generalised intuitionistic fuzzy soft set** over (U, E).

Definition 1.16 Let U be an initial universe and E be a set of parameters. Let IVF^U be the set of all interval-valued fuzzy subsets of U and $A \subseteq E$. Then, the pair (F, A) is called an ***interval-valued fuzzy soft set*** over U, where F is a mapping given by $F: A \to \mathrm{IVF}^U$.

Definition 1.17 Let U be an initial universe and E be a set of parameters. Let IVIFS^U be the set of all interval-valued intuitionistic fuzzy soft sets on U and $A \subseteq E$. Then, the pair (F, A) is called an ***interval-valued intuitionistic fuzzy soft set*** over U, where F is a mapping given by $F: A \to \mathrm{IVIFS}^U$.

For any parameter $\varepsilon \in A$, $F(\varepsilon)$ is referred as the interval intuitionistic fuzzy value set of parameter of ε. It is actually an interval-valued intuitionistic fuzzy set of U where $x \in U$ and $\varepsilon \in A$. Then, $F(\varepsilon)$ can be written as $F(\varepsilon) = \{(x, \mu_{F(\varepsilon)}(x), \gamma_{F(\varepsilon)}(x)): x \in U\}$. Here, $\mu_{F(\varepsilon)}(x)$ is the interval-valued fuzzy membership degree that object x holds parameter ε and $\gamma_{F(\varepsilon)}(x)$ is the interval-valued fuzzy membership degree that object x does not hold parameter ε.

Definition 1.18 A *comparison table* is a square table in which the number of rows and the number of columns are equal and both are labelled by the object name of the universe such as $h_1, h_2, h_3, \ldots, h_n$ and the entries are c_{ij}, where c_{ij} = the number of parameters for which the value of h_i exceeds or is equal to the value h_j.

Definition 1.19 The *row sum* of an object h_i is denoted by r_i and is calculated by using the formula

$$r_i = \sum_{j=1}^{n} c_{ij}.$$

The *column sum* of an object h_j is denoted by t_j and is calculated by using the formula

$$t_j = \sum_{i=1}^{n} c_{ij}.$$

Definition 1.20 The *score* of an object h_i is denoted by S_i and is calculated by using the formula

$$S_i = r_i - t_j.$$

1.4 Fuzzy Topological Space

The notion of fuzzy topology was introduced by C.L. Chang in 1968 [1]. It is the extension of the concepts of ordinary topological space. Let X be a non-empty set and $I = [0, 1]$ be the unit closed interval. For X, I^X denotes the collection of all mappings from X into I. A member λ of I^X is called a fuzzy set. The union $(\vee \lambda_i)$ the intersection, $(\wedge \lambda_i)$ of a family $\{\lambda_i\}$ of fuzzy sets of X is defined to be the mapping $\sup \lambda_i$ (inf λ_i). For any two members λ and β of I^X, $\lambda \geq \beta$ if and only if $\lambda(x) \geq \beta(x)$ for each $x \in X$. 0 and 1 denotes the constant mappings family whole of X to 0 and 1, respectively. The complement λ^c of a fuzzy set λ of X is $1 - \lambda$ defined as $(1 - \lambda)(x) = 1 - \lambda(x)$ for each $x \in X$. If λ is a fuzzy set of X and β is a fuzzy set of Y, then $\lambda \times \beta$ is a fuzzy set of $X \times Y$ defined by $(\lambda \times \beta)(x, y) = \min \{\lambda(x), \beta(y)\}$ for each $(x, y) \in X \times Y$.

1.21 A fuzzy set in X is called a fuzzy point iff it takes the value 0 for all $y \in X$ except one say $x \in X$. If its value at x is $p(0 < p \leq 1)$, we denote the fuzzy point by x_p where the point x is called its support and p its value.

Or equivalently, a fuzzy point xp in X is a special fuzzy set with membership function denoted by $x_p(y) = p$, $x = y(0 \leq p \leq 1) = 0$, $x \neq y$.

Note

(i) Let α be a fuzzy set in X, and then, $x_p \subseteq \alpha$ implies $p \leq \alpha(x)$. In particular, $x_p \subseteq y_q$ implies and implied by $x = y$, $p \leq q$
(ii) $x_p \in \alpha$ implies and implied by $p \leq \alpha(x)$.

1.22

(i) A fuzzy point x_p is said to be quasi-coincident (q coincident) with a fuzzy subset α, denoted by $x_p q \alpha$ iff $p + \alpha(x) > 1$ or $p > 1 - \alpha(x)$ or $p > \alpha^c(x)$.
(ii) A fuzzy subset α is q coincident with another fuzzy subset β denoted by $\alpha q \beta$ iff there exists $x \in X$ such that $\alpha(x) + \beta(x) > 1$, $\mu_\alpha(x) > 1 - \beta(x)$, or $\alpha(x) > \beta^c(x)$. In this case, we say that two fuzzy subsets α and β are q coincident (with each other) at x. It is clear that if α and β are q coincident at x, then both $\alpha(x)$ and $\beta(x)$ are not zero, and hence, α and β intersect at x. If α does not q coincident with β, then we write $\alpha/q\beta$.

Note $\alpha \leq \beta$ iff α and β^c are not q coincident. In particular, $x_p \in \alpha$ iff x_p is not q coincident with α^c. It follows from the fact $\alpha(x) \leq \beta(x)$ implies and implied by $\alpha(x) + \beta^c(x) \leq \beta(x) + \beta^c(x) = 1$, i.e. $x_p \, q/\alpha^c$.

1.23 Let $f: X \rightarrow Y$ be a mapping. If λ be a fuzzy set of X, we define $f(\lambda)$ as

$$f(\lambda)(y) = \sup_{x \in f^{-1}(y)} \lambda(x) \quad \text{if} \quad f^{-1}(y) \neq \phi$$

$$= 0$$

for each $y \in Y$, and if β is a fuzzy set of Y, we define $f^{-1}(\beta)$ as $f^{-1}(\beta)(x) = \beta f(x)$ for each $x \in X$.

Remark

(i) $f^{-1}(\beta^c) = (f(\beta))^c$ for any fuzzy subset β of Y
(ii) $f(f^{-1}(\beta)) \leq \beta$ for any fuzzy subset β of Y
(iii) $\lambda \leq f^{-1}(f(\lambda))$ for any fuzzy subset λ of X.

1.24

(a) The product $f_1 \times f_2 : X_1 \times X_2 \to Y_1 \times Y_2$ of mappings $f_1 : X_1 \to Y_1$ and $f_2 : X_2 \to Y_2$ is defined by $f_1 \times f_2(x_1, x_2) = (f(x_1), f(x_2))$ for each $(x_1, x_2) \in X_1 \times X_2$ and
(b) for a mapping $f : X \to Y$, the graph $g : X \to X \times Y$ of f is defined by $g(x) = (x, f(x))$ for each $x \in X$.

Result 1.25 Let $f : X \to Y$ be a mapping and $\{\beta_j\}$ be a family of fuzzy sets of Y, and then

(a) $f^{-1}(\vee\{\beta_j\}) = \vee f^{-1}(\{\beta_j\})$
(b) $f^{-1}(\wedge\{\beta_j\}) = \wedge f^{-1}(\{\beta_j\})$

Lemma 1.26 *If λ be a fuzzy set of X and β be a fuzzy set of Y, then $1 - (\lambda \times \beta) = (\lambda^c \times 1) \vee (1 \times \beta^c)$.*

Proof We know $(\lambda \times \beta)(x, y) = \min(\lambda(x), \beta(y))$ for each $(x, y) \in X \times Y$. Therefore,

$$(1 - (\lambda \times \beta))(x, y) = 1 - (\lambda \times \beta)(x, y) = \max\{(1 - \lambda(x)), 1 - \beta(y)\}$$
$$= \max\{(\lambda^c \times 1)(x, y), (1 \times \beta^c)(x, y)\}$$
$$= (\lambda^c \times 1)V(1 \times \beta^c) \text{ for each}(x, y) \in X \times Y$$

$\square$

Lemma 1.27 *For mapping $f_i : X_i \to Y_i$ and fuzzy subsets X_i of Y_i, $i = 1, 2$, we have*

$$(f_1 \times f_2)^{-1}(\lambda_1 \times \lambda_2)(x_1, x_2) = (\lambda_1 \times \lambda_2)(f_1(x_1), f_2(x_2)) = \min\{\lambda_1 f_1(x_1), \lambda_2 f_2(x_2)\}$$
$$= \min\left\{(f_1^{-1}(\lambda_1)(x_1), (f_2)^{-1}(\lambda_2)(x_2)\right\}$$
$$= \left(f_1^{-1}(\lambda_1) \times (f_2)^{-1}(\lambda_2)\right)(x_1, x_2)$$

Therefore, $(f_1 \times f_2)^{-1}(\lambda_1 \times \lambda_2) = f_1^{-1}(\lambda_1) \times (f_2)^{-1}(\lambda_2)$.

Lemma 1.28 *Let $g : X \to X \times Y$ be the graph of a mapping $f : X \to Y$. Then, if λ is a fuzzy subset of X and β be a fuzzy subset of Y, $g^{-1}(\lambda \times \beta) = \lambda \cap f^{-1}(\beta)$.*

Proof For each $x \in X$, we have $g^{-1}(\lambda \times \beta)(x) = (\lambda \times \beta)g(x) = (\lambda \times \beta)(x, f(x)) = \min(\lambda(x), \beta f(x)) = (\lambda \cap f^{-1}(\beta))(x)$.

Therefore, $g^{-1}(\lambda \times \beta) = \lambda \cap f^{-1}(\beta)$.

The notion of a fuzzy topology was introduced by C.L. Chang in 1968 [1]. It is an extension of the concept of ordinary topology. $\square$

Definition 1.29 A family $F \subseteq I^X$ of fuzzy subsets is called a fuzzy topology (in the sense of Chang) for X if it satisfies the following:

(i) $0 = \mu_\phi$ and $1 = \mu_X \in F$
(ii) for all $\alpha, \beta \in F$ implies $\alpha \wedge \beta \in F$
(iii) If $\alpha_j \in F$ for each j, then $\sup_{j \in \Lambda} \alpha_j \in F$

Then, F is called a fuzzy topology for X, and the pair (X, F) is called a fuzzy topological space. Members of F are called fuzzy topological space. Members of F are called fuzzy open subsets and their pseudo-complement $\alpha^c = 1 - \alpha$ fuzzy closed subsets. In 1976, R. Lowen introduced a new definition of fuzzy topology replacing the condition (i) by (i') for all a (Constant function) $a \in F$.

Both the definitions are used by fuzzy topologists. One of the main advantages of Lowen's definition over Chang's definition is that all constant functions are fuzzy continuous.

Definition 1.30 Let α be a fuzzy subset in a fuzzy topological space (ft space) (X, F), and then

(a)   The closure of α denoted by $Cl(\alpha)$ is defined by $Cl(\alpha) = \inf \{\beta: \beta \supseteq \alpha, \beta^c \in F\}$
(b)   The interior of α is denoted by $\alpha^0 = \text{int } \alpha = \sup \{\beta: \beta \subseteq \alpha, \beta \in F\}$.

Definition 1.31 Base and Sub-base for fuzzy topology
Let F be a fuzzy topology. A subfamily ß of F is a base for F iff each member of F can be expressed as the union of some members of ß. A subfamily S of F is a sub-base of F iff the family of finite intersection of members of S forms a base for F. Let (X, F) be an fts. A subfamily ßoF is a base for F iff for each $\alpha \in F$, there exists $(\alpha_j)_{j\in\Lambda} \subset$ ß such that $\alpha = \vee_{j\in\Lambda}\alpha_j$, a subfamily S of F is called a sub-base of F iff the family of finite infima of members of S is a base for F, (X, F) is said to satisfy the second axiom of countability or is said to be C_{11} space iff F has a countable base.

Example 1.32 Let α_1, α_2, and α_3 be fuzzy subsets of $I = [0, 1]$ defined as

$$\begin{aligned}
\alpha_1(x) &= 0, \quad 0 \leq x \leq 1/2 \\
&= 2x - 1, \quad 1/2 \leq x \leq 1 \\
\alpha_2(x) &= 1, \quad 0 \leq x \leq 1/4 \\
&= -4x + 2, \quad 1/4 \leq x \leq 1/2 \\
&= 0, \quad 1/2 \leq x \leq 1 \\
\alpha_3(x) &= 0, \quad 0 \leq x \leq 1/4 \\
&= 4x - 1, \quad 1/4 \leq x \leq 1
\end{aligned}$$

Then, $F = \{0, \alpha_1, \alpha_2, \alpha_1 \cup \alpha_2, 1\}$ is a fuzzy topology on I.

By easy computations, it can be seen that $Cl(\alpha_1) = \alpha_2^c$, $Cl(\alpha_2) = \alpha_1^c$, $Cl(\alpha_1 \cup \alpha_2) = 1$, $\text{int}(\alpha_1^c) = \alpha_2$, $\text{Int}(\alpha_2^c) = \alpha_1$, $\text{int}(\alpha_1 \cup \alpha_2)^c = 0$.

Note For a fuzzy set λ of a fuzzy space (X, F), (a) $1 - \text{int}\lambda = Cl(1 - \lambda)$ and (b) $1 - Cl\lambda = \text{int}(1 - \lambda)$.

Definition 1.33 A fuzzy set α in (X, F) is called a neighbourhood of fuzzy point p and x_p iff there exists a $\beta \in F$ such that $x_p \in \beta \leq \alpha$ and a neighbourhood α is said to be open neighbourhood iff α is a fuzzy open. The family consisting of all the neighbourhoods of x_p is called the system of neighbourhoods of x_p.

Result 1.34 Let f be a function from X to Y. Let x_p be a fuzzy point Y of X be a fuzzy subset of X and β be a fuzzy subset in Y then

   (i) if $f(x_p)$ q β, then $x_p q f^{-1}(\beta)$
   (ii) if $x_p q \alpha$, then $f(x_p)$ q $f(\alpha)$
   (iii) $x_p \in f^{-1}(\beta)$ if $f(x_p) \in \beta$
   (iv) $f(x_p) \in f(\alpha)$ if $x_p \in \alpha$.

Definition 1.35 A fuzzy subset α in (X, F) is called q neighbourhood of x_p iff there exists a $\beta \in F$ such that x_p q $\beta \leq \alpha$. The family consisting of all the q neighbourhoods of x_p is called the system of q neighbourhoods of x_p.

In point set theory, it is well known that the closure of the product of the sets is the product of closures, i.e. $Cl(\alpha \times \beta) = Cl\alpha \times Cl\beta$. But it is not true in fuzzy settings, i.e. $Cl(\gamma \times \rho) \neq Cl\gamma \times Cl\rho$ where γ and ρ are two fuzzy subsets.

Let $X = Y = I$. Consider the fuzzy sets λ, μ, γ, and ρ of I defined as

$$
\begin{aligned}
\lambda(x) \quad &= -4/3x + 1, \quad 0 \leq x \leq 3/4 \\
&= 0, \quad\quad\quad\quad 3/4 \leq x \leq 1 \\
\mu(x) \quad &= 0, \quad\quad\quad\quad 0 \leq x \leq 3/4 \\
&= 4x - 3, \quad\quad 3/4 \leq x \leq 1 \\
\gamma(x) \quad &= 1/6, \quad\quad\quad \text{if } x \leq 2/3 \\
&= 0, \quad\quad\quad\quad \text{otherwise} \\
\rho(x) \quad &= 2/5, \quad\quad\quad \text{if } x \leq 4/5 \\
&= 0, \quad\quad\quad\quad \text{otherwise}
\end{aligned}
$$

Let $F_X = \{0, \lambda', 1\}$ and $F_Y = \{0, \mu', 1\}$. Clearly, F_X and F_Y are fuzzy topological spaces. We observe $\lambda \not\geq \gamma$ (put $x = 2/5$ in $\lambda(x) = -4/3x + 1 = -1/9 \geq \gamma$.

$\mu \not\geq \rho$ (put $x = 4/5$ in $\mu(x) = 4x - 3 = 1/5 \geq \rho$) $Cl\gamma = 1$ in X, $Cl\rho = 1$ in Y and $Cl\gamma \times Cl\rho = 1$. Also, $\lambda \times 1 \cdot 1 \times \mu$ and $(\lambda \times 1) \vee (1 \times \mu)$ are fuzzy closed sets of $X \times Y$ [Th: If λ be a fuzzy closed set of X and μ be a fuzzy closed set of Y, then $\lambda \times \mu$ is a fuzzy closed set of $X \times Y$].

Now, (I) $\gamma \times \rho$ is defined by

$$
\begin{aligned}
(\gamma \times \rho)(x, y) &= \min(\gamma(x), \rho(y)) \quad \text{if } (x, y) = (2/5, 4/5) \\
&= \min(1/6, 2/5) = 1/6 \\
&= 0 \text{ otherwise}
\end{aligned}
$$

and (II) $((\lambda \times 1) \vee (1 \times \mu))(2/3, 4/5) = \max(1/9, 1/5) = 1/5$ implies $(\lambda \times 1) \vee (1 \times \mu) \geq \gamma \times \rho$, and hence, $1 \neq (\lambda \times 1) \vee (1 \times \mu) \geq \mathrm{Cl}(\gamma \times \rho)$ shows that $\mathrm{Cl}(\gamma \times \rho) \neq \mathrm{Cl}\gamma \times \mathrm{Cl}\rho$.

Note If λ be a fuzzy set of a fuzzy space X and μ be a fuzzy set of a fuzzy space Y, then

(i) $\mathrm{Cl}\lambda \times \mathrm{Cl}\mu \geq \mathrm{Cl}(\lambda \times \mu)$
(ii) $\mathrm{int}\,\lambda \times \mathrm{int}\,\mu \leq \mathrm{int}(\lambda \times \mu)$.

1.5 Intuitionistic Fuzzy Topological Space

In 1983, Atanassov introduced the concept of 'intuitionistic fuzzy set'. Using this type of generalised fuzzy set, Coker defined 'Intuitionistic fuzzy topological spaces' [3]. In 1996, Coker and Demirci introduced the basic definitions and properties of intuitionistic fuzzy topological spaces in ˇSostak's sense, which is a generalised form of 'fuzzy topological spaces' developed by ˇSostak.

Definition 1.36 Let X be a non-empty fixed set and I the closed unit interval $[0, 1]$. An intuitionistic fuzzy set (IFS) A is an object having the form

$$A = \{\langle x, \mu_A(x), \nu_A(x)\rangle : x \in X\}$$

where the mappings $\mu_A \colon X \to I$ and $\nu_A \colon X \to I$ denote the degree of membership (namely, $\mu_A(x)$) and the degree of non-membership (namely, $\nu_A(x)$) of each element $x \in X$ to the set A, respectively, and $0 \leq \mu_A(x) + \nu_A(x) \leq 1$ for each $x \in X$. The complement of the IFS A is $\bar{A} = \{\langle x, \nu_A(x), \mu_A(x)\rangle : x \in X\}$. Obviously, every fuzzy set A on a non-empty set X is an IFS having the form $A = \{\langle x, \mu_A(x), 1 - \mu_A(x)\rangle : x \in X\}$.

For a given non-empty set X, we denote the family of all IFSs in X by the symbol ζ^X.

Definition 1.37 Let X be a non-empty set and $x \in X$ a fixed element in X. If $r \in I_0$, $s \in I_1$ are fixed real numbers such that $r + s \leq 1$, then the IFS $x_{r,s} = \langle y, x_r, 1 - x_{1-s}\rangle$ is called an intuitionistic fuzzy point (IFP) in X, where r denotes the degree of membership of $x_{r,s}$, s the degree of non-membership of $x_{r,s}$, and $x \in X$ the support of $x_{r,s}$. The IFP $x_{r,s}$ is contained in the IFS A ($x_{r,s} \in A$) if and only if $r\langle \mu_A(x), s\rangle \gamma_A(x)$.

Definition 1.38

(i)   An IFP $x_{r,s}$ in X is said to be quasi-coincident with the IFS A, denoted by $x_{r,s}qA$, if and only if $r > \gamma A(x)$ or $s < \mu A(x)$. $x_{r,s}qA$ if and only if $x_{r,s} \notin A$.

(ii) The IFSs A and B are said to be quasi-coincident, denoted by AqB if and only if there exists an element $x \in X$ such that $\mu_A(x) > \gamma_B(x)$ or $\gamma_A(x) < \mu_B(x)$. If A is not quasi-coincident with A, we denote $A\not qB$. AqB if and only if $A \subseteq B$.

Definition 1.39 Let a and b be two real numbers in $[0, 1]$ satisfying the inequality $a + b \leq 1$. Then, the pair $\langle a, b \rangle$ is called an intuitionistic fuzzy pair.

Let $\langle a_1, b_1 \rangle, \langle a_2, b_2 \rangle$ be two intuitionistic fuzzy pairs. Then, define

(i) $\langle a_1, b_1 \rangle \leq \langle a_2, b_2 \rangle$ if and only if $a_1 \leq a_2$ and $b_1 \geq b_2$;
(ii) $\langle a_1, b_1 \rangle = \langle a_2, b_2 \rangle$ if and only if $a_1 = a_2$ and $b_1 = b_2$;
(iii) if $\{\langle a_i, b_i \rangle : i \in J\}$ is a family of intuitionistic fuzzy pairs, then $\vee \langle a_i, b_i \rangle = \langle \vee a_i, \wedge b_i \rangle$ and $\wedge \langle a_i, b_i \rangle = \langle \wedge a_i, \vee b_i \rangle$;
(iv) the complement of an intuitionistic fuzzy pair $\langle a, b \rangle$ is the intuitionistic fuzzy pair defined by $\overline{\langle a, b \rangle} = \langle b, a \rangle$;
(v) $1_\sim = \langle 1, 0 \rangle$ and $0_\sim = \langle 0, 1 \rangle$.

Definition 1.40 An intuitionistic fuzzy topology (IFT) in Chang's sense on a non-empty set X is a family τ of IFSs in X satisfying the following axioms:

$(T1) 0_\sim, 1_\sim \in \tau$, where $0_\sim = \{\langle x, 0, 1 \rangle : x \in X\}$ and $1_\sim = \{\langle x, 1, 0 \rangle : x \in X\}$;
$(T2)$ $G_1 \wedge G_2 \in \tau$ for any $G_1, G_2 \in \tau$;
$(T3)$ $\vee G_i \in \tau$ for any arbitrary family $\{G_i : i \in J\} \subseteq \tau$.

In this case, the pair (X, τ) is called Chang intuitionistic fuzzy topological space and each IFS in τ is known as intuitionistic fuzzy open set in X.

Definition 1.41 An IFS ξ on the set ζ^X is called an intuitionistic fuzzy family (IFF) in X. In symbols, we denote such an IFF in the form $\xi = \langle \mu_\xi, \nu_\xi \rangle$.

Let ξ be an IFF on X. Then, the complemented IFF of ξ on X is defined by $\xi^* = \langle \mu_{\xi^*}, \nu_{\xi^*} \rangle$, where $\mu_{\xi^*}(A) = \mu_\xi(\overline{A})$ and $\nu_{\xi^*}(A) = \nu_\xi(\overline{A})$, for each $A \in \zeta^X$. If τ is an IFF on X, then for any $A \in \zeta^X$, we construct the intuitionistic fuzzy pair $\langle \mu_\tau(A), \nu_\tau(A) \rangle$ and use the symbol $\tau(A) = \langle \mu_\tau(A), \nu_\tau(A) \rangle$.

Definition 1.42 An IFT in ˇSostak's sense on a non-empty set X is an IFT τ on X satisfying the following axioms:

$(T1) \tau(0_\sim) = \tau(1_\sim) = 1_\sim$;
$(T2) \tau(A \cap B) \geq \tau(A) \wedge \tau(B)$ for any $A, B \in \zeta^X$;
$(T3) \tau(\cup A_i) \geq \wedge \tau(A_i)$ for any $\{A_i : i \in J\} \subseteq \zeta^X$.

In this case, the pair (X, τ) is called an intuitionistic fuzzy topological space in ˇSostak's sense (IFTS). For any $A \in \zeta^X$, the number $\mu_\tau(A)$ is called the openness degree of A, while $\nu_\tau(A)$ is called the non-openness degree of A.

1.6 Relation Among Fuzzy Set, Rough Set, Soft Set, and Their Generalisations

1.6.1 Relation Between Fuzzy Set and Soft Set

Theorem 1.43 *Every fuzzy set may be considered a soft set.*

In order to better understand the relationship, let us consider the following example.

Suppose that there are six alternatives in the universe of houses $U = \{h_1, h_2, h_3, h_4, h_5, h_6\}$ and we consider the single parameter 'quality of the houses' to be a linguistic variable. For this variable, we define the set of linguistic terms T (quality) = {best, good, fair, poor}. Each linguistic term is associated with its own fuzzy set. Let us consider two of them as follows:

$F_{[\text{best}]} = \{(h_1, 0.25), (h_2, 0.6), (h_5, 0.9), (h_6, 1.0)\}$ and
$F_{[\text{poor}]} = \{(h_1, 0.9), (h_2, 0.4), (h_3, 1.0), (h_4, 1.0), (h_5, 0.25)\}$.
Now the α-level sets of $F_{[\text{poor}]}$ are
$F_{[\text{poor}]}(0.25) = \{h_1, h_2, h_3, h_4, h_5\}$.
$F_{[\text{poor}]}(0.4) = \{h_1, h_2, h_3, h_4\}$.
$F_{[\text{poor}]}(0.9) = \{h_1, h_3, h_4\}$.
$F_{[\text{poor}]}(1.0) = \{h_3, h_4\}$.

The values $A = \{0.25, 0.4, 0.9, 1.0\} \subset [0.1]$ can be treated as a set of parameters, such that the mapping $F_{[\text{poor}]}: A \rightarrow P(U)$ gives the approximate value set $F_{[\text{poor}]}(\alpha)$ for $\alpha \in A$. Thus, we can write the equivalent soft set—$(F_{[\text{poor}]}, [0,1]) = \{(0.25, \{h_1, h_2, h_3, h_4, h_5\}), (0.4, \{h_1, h_2, h_3, h_4\}), (0.9, \{h_1, h_3, h_4\}), (1.0, \{h_3, h_4\})\}$.

1.6.2 Relation Between Rough Set and Soft Set

Theorem 1.44 *Every rough set may be considered a soft set.*

In order to better understand the relationship, let us consider the following example.

Suppose that a subset of five houses $X = \{h_1, h_2, h_3, h_4, h_5\}$ in the universe of $U = \{h_1, h_2, h_3, h_4, h_5, h_6\}$ are under consideration. We construct the information table:

House	h_1,	h_2,	h_3,	h_4,	h_5,	h_6,
Quality	Fair	Fair	Best	Good	Best	Fair
Price	Cheap	Cheap	Middle	Expensive	Middle	Cheap
Place	City	City	Village	City	Village	City

The rows of the table are leveled by attributes, and the table entries are the attribute values for each home.

Each column in the table can thus be seen as summarising the available information on a specific home. The table evaluates all six houses in terms of three attributes, 'quality', 'price', and 'place". These three attributes are characterised by the value sets {best, good, fair, poor},{expensive, middle, cheap},{village, city}, respectively.

Now, the equivalence classes are

$$[h_1]_R = [h_2]_R = [h_6]_R = \{h_1, h_2, h_6\}.$$
$$[h_3]_R = [h_5]_R = \{h_3, h_5\}.$$
$$[h_4]_R = \{h_4\}.$$

Thus,

$$R_*(X) = \{h_3, h_4, h_5\} \text{ and } R^*(X) = \{h_1, h_2, h_3, h_4, h_5, h_6\}$$
$$R(X) = \{\{h_3, h_4, h_5\}, \{h_1, h_2, h_3, h_4, h_5, h_6\}\}.$$

Thus, every rough set $R(X)$ of X may also be considered a soft set with the representation

$$(F, E) = \{(p_1(x), R_*(X)), (p_2(x), R^*(X))\}.$$

Nanda and Majumdar [6] introduced the notion of fuzzy rough sets. In 1998, Chakrabarty et al. [2] approached intuitionistic fuzzy rough sets (IF rough set), they constructed an IF rough set (A, B) of the rough set (P, Q), where A and B are both IF sets in X such that $A \subseteq B$ i.e. $\mu_A \leq \mu_B$ and $v_A \geq v_B$. From this point of view, the lower approximation A and the upper approximation B are both IF sets. Jena and Ghosh [4] reintroduced the same notion. Samanta and Mondal [9] also introduced this notion, but they called it a rough IF set. They also defined the concept of IF rough set. According to them, an IF rough set is a couple (A, B) such that A and B are both fuzzy rough sets (in the sense of Nanda and Majumdar) and A is included in the complement of B. According to Samanta and Mondal (2001), an intuitionistic fuzzy rough set (A, B) is a generalisation of an IF set in which membership and non-membership functions are no longer fuzzy sets but fuzzy rough sets A and B. On the other hand, for Chakrabarty et al., an intuitionistic fuzzy rough set (A, B) is a generalisation of a fuzzy rough set in which upper and lower approximations are no longer fuzzy sets but IF sets A and B. Rizvi et al. [8] described their proposal as 'Rough intuitionistic fuzzy set' in which the lower and upper approximations themselves are not intuitionistic fuzzy sets in X but intuitionistic fuzzy sets in the class of equivalence classes.

1.7 Soft Multi-Sets and Their Basic Properties

Let $\{U_i : i \in I\}$ be a collection of universes such that $\bigcap_{i \in I} U_i = \phi$ and let $\{E_{U_i} : i \in I\}$ be a collection of sets of parameters. Let $U = \prod_{i \in I} P(U_i)$ where $P(U_i)$ denotes the power set of U_i, $E = \prod_{i \in I} E_{U_i}$. The set of all soft multi-set over (U, E) is denoted by $SMS(U, E)$.

Definition 1.45 A soft multi-set $(F, E) \in SMS(U, E)$ is called a null soft multi-set denoted by $\tilde{\phi}$, if for all $e \in E$, $F(e) = \phi$.

Definition 1.46 A soft multi-set $(F, E) \in SMS(U, E)$ is called an absolute soft multi-set denoted by $\tilde{E}$, if for all $e \in E$, $F(e) = U$.

Definition 1.47 The relative complement of a soft multi-set (F, A) over (U, E) is denoted by $(F, A)'$ and is defined by $(F, A)' = (F', A)$, where $F' : A \to U$ is a mapping given by $F'(e) = U - F(e), \forall e \in E$.

Example 1.48 Let us consider that there are three universes U_1, U_2, and U_3.
Let $U_1 = \{h_1, h_2, h_3, h_4\}$, $U_2 = \{c_1, c_2, c_3\}$, and $U_3 = \{v_1, v_2\}$. Let $\{E_{U_1}, E_{U_2}, E_{U_3}\}$ be a collection of sets of decision parameters related to the above universes, where $E_{U_1} = \{e_{U_1,1}, e_{U_1,2}, e_{U_1,3}\}, E_{U_2} = \{e_{U_2,1}, e_{U_2,2}, e_{U_2,3}\}, E_{U_3} = \{e_{U_3,1}, e_{U_3,2}, e_{U_3,3}\}$.

Let $U = \prod_{i=1}^{3} P(U_i)$, $E = \prod_{i=1}^{3} E_{U_i}$ and

$$A = \{e_1 = (e_{U_1,1}, e_{U_2,1}, e_{U_3,1}), e_2 = (e_{U_1,1}, e_{U_2,2}, e_{U_3,1})\}$$

Then, the relative complement of the soft multi-set

$$(F, A) = \{(e_1, (\{h_1, h_2\}, \{c_1, c_2\}, \{v_1\})), (e_2, (\{h_3, h_4\}, \{c_1, c_3\}, \{v_2\}))\},$$

is $(F, A)' = \{(e_1, (\{h_3, h_4\}, \{c_3\}, \{v_2\})), (e_2, (\{h_1, h_2\}, \{c_2\}, \{v_1\}))\}$. Clearly, $\tilde{\phi}' = \tilde{E}$ and $\tilde{E}' = \tilde{\phi}$.

Proposition 1.49 *If (F, A) and (G, B) are two soft multi-sets over (U, E), then we have the following:*

(i) $((F, A) \; \tilde{\cup} \; (G, B))' = (F, A)' \; \tilde{\cap} \; (G, B)'$
(ii) $((F, A) \; \tilde{\cap} \; (G, B))' = (F, A)' \; \tilde{\cup} \; (G, B)'$

Proof Straight forward.   □

Definition 1.50 A soft multi-set $(F, A) \in SMS(U, E)$ is called a soft multi-point in (U, E), denoted by $e_{(F,A)}$, if for the element $e \in A$, $F(e) \neq \varphi$ and $\forall e' \in A - \{e\}$, $F(e') = \varphi$.

Example 1.51 Let us consider that there are three universes U_1, U_2, and U_3.

Let $U_1 = \{h_1, h_2, h_3, h_4\}$, $U_2 = \{c_1, c_2, c_3\}$, and $U_3 = \{v_1, v_2\}$. Let $\{E_{U_1}, E_{U_2}, E_{U_3}\}$ be a collection of sets of decision parameters related to the above universes, where

$$E_{U_1} = \{e_{U_1,1} = \text{expensive}, e_{U_1,2} = \text{cheap}, e_{U_1,3} = \text{wooden}\}$$
$$E_{U_2} = \{e_{U_2,1} = \text{expensive}, e_{U_2,2} = \text{cheap}, e_{U_2,3} = \text{sporty}\},$$
$$E_{U_3} = \{e_{U_3,1} = \text{expensive}, e_{U_3,2} = \text{cheap}, e_{U_3,3} = \text{in Kuala Lumpur}\}.$$

Let $U = \prod_{i=1}^{3} P(U_i)$, $E = \prod_{i=1}^{3} E_{U_i}$ and

$$A = \{e_1 = (e_{U_1,1}, e_{U_2,1}, e_{U_3,1}), e_2 = (e_{U_1,1}, e_{U_2,2}, e_{U_3,1}), e_3 = (e_{U_1,2}, e_{U_2,3}, e_{U_3,1})\}$$

Then, the soft multi-set $(F, A) = \{(e_1, (\{h_1, h_2\}, \{c_1, c_2\}, \phi))\}$ is the soft multi-point, and it is denoted by $e_{1(F,A)}$.

Definition 1.52 A soft multi-point $e_{(F,A)}$ is said to be in the soft multi-set (G, B), denoted by $e_{(F,A)} \tilde{\in} (G, B)$, if $(F, A) \tilde{\subseteq} (G, B)$.

Example 1.53 The soft multi-point $e_{1(F,A)}$ as in Example 1.51, in the soft multi-set $(G, B) = \{(e_1, (\{h_1, h_2\}, \{c_1, c_2\}, \{v_1\})), (e_2, (\{h_3, h_4\}, \{c_1, c_3\}, \{v_2\})), (e_3, (\{h_1, h_3, h_4\}, \{c_1, c_3\}, \{v_1, v_2\}))\}$, i.e. $e_{(F,A)} \tilde{\in} (G, B)$.

Proposition 1.54 *Let $e_{(F,A)}$ be a soft multi-point and (G, B) be the soft multi-set in* $\mathrm{SMS}(U, E)$. *If $e_{(F,A)} \tilde{\in} (G, B)$, then $e_{(F,A)} \tilde{\notin} (G, B)^c$.*

Proof If $e_{(F,A)} \tilde{\in} (G, B)$, then $(F, A) \tilde{\subseteq} (G, B)$, i.e. for the element $e \in A, F(e) \subseteq G(e)$. This implies $F(e) \not\subseteq U - G(e) = G^c(e)$, i.e. $(F, A) \not\tilde{\subseteq} (G, B)^c$. Therefore, we have $e_{(F,A)} \tilde{\notin} (G, B)^c$. $\qquad\square$

Remark 1.55 The converse of the above proposition is not true in general.

Example 1.56 If we consider the soft multi-point,

$$e_{1(F,A)} = \{(e_1, (\{h_1, h_2\}, \{c_1, c_2\}, \phi))\}$$

as in Example 1.51 and a soft multi-set

$$(G, B) = \{(e_1, (\{h_1, h_3\}, \{c_2, c_3\}, \{v_1\})), (e_2, (\{h_2, h_4\}, \{c_1, c_3\}, \{v_2\})), (e_3, (\{h_4\}, \{c_1\}, \{v_2\}))\}.$$

Then, $e_{1(F,A)} \tilde{\notin} (G, B)$ and also $e_{1(F,A)} \tilde{\notin} (G, B)^c$

$$= \{(e_1, (\{h_2, h_4\}, \{c_1\}, \{v_2\})), (e_2, (\{h_1, h_3\}, \{c_2\}, \{v_1\})), (e_3, (\{h_1, h_2, h_3\}, \{c_2, c_3\}, \{v_1\}))\}.$$

Definition 1.57 Let $(F, A) \in \mathrm{SMS}(U, E)$ and $x \in U_i$, for some i. Then, we say that $x \in (F, A)$ and read as x belongs to the soft multi-set (F, A) if $x \in F_{e_{U_i,j}}, \forall j$.

Example 1.58 Let us consider the soft multi-set $(F, A) = \{(e_1, (\{h_1, h_2\}, \{c_1, c_2\},$ $\{v_1\})), (e_2, (\{h_3, h_4\}, \{c_1, c_3\}, \{v_2\}))\}$, as in Example 1.53, and then, for the element $c_1 \in U_2$, we say that $c_1 \in (F, A)$, since $c_1 \in F_{e_{U_{2,1}}} = \{c_1, c_2\}$ and $c_1 \in F_{e_{U_{2,2}}} = \{c_1, c_3\}$, but $h_1, h_2 \notin (F, A)$ since $h_1, h_2 \in F_{e_{U_{1,1}}} = \{h_1, h_2\}$ but $h_1, h_2 \notin F_{e_{U_{1,2}}} = \{h_3, h_4\}$.

Remark 1.59 For any $x \in U_i$, we say that $x \notin (F, A)$ if $x \notin F(e_{U_{i,j}})$ for some $e_{U_{i,j}} \in a_k$, $a_k \in A$.

1.8 Soft Multi-Topological Spaces

Recently, D. Tokat, I. Osmanoglu and also Mukherjee and Das introduced soft multi-topology. In this section, the notion of relative complement of soft multi-set, soft multi-point, soft multi-set topology, soft multi-closed set, soft multi-basis, soft multi-sub-basis, neighbourhood and neighbourhood system, interior and closure of a soft multi-set, etc. are to be introduced and their basic properties are investigated. It is seen that a soft multi-topological space gives a parameterised family of topological spaces.

Definition 1.60 A subfamily τ of $SMS(U, E)$ is called soft multi-set topology on (U, E), if the following axioms are satisfied:

$[O_1]$ $\tilde{\phi}, \tilde{E} \in \tau$,
$[O_2]$ $\{(F^k, A^k) | k \in K\} \subseteq \tau \Rightarrow \tilde{\cup}_{k \in K}(F^k, A^k) \in \tau$,
$[O_3]$ If $(F, A), (G, B) \in \tau$, then $(F, A) \tilde{\cap} (G, B) \in \tau$.

Then, the pair $((U, E), \tau)$ is called soft multi-topological space. The members of τ are called soft multi-open sets (or τ—open soft multi-sets or simply open sets), and the conditions $[O_1]$, $[O_2]$, and $[O_3]$ are called the axioms for soft multi-open sets.

Example 1.61 Let us consider that there are three universes U_1, U_2, and U_3. Let $U_1 = \{h_1, h_2, h_3, h_4\}$, $U_2 = \{c_1, c_2, c_3\}$, and $U_3 = \{v_1, v_2\}$. Let $\{E_{U_1}, E_{U_2}, E_{U_3}\}$ be a collection of sets of decision parameters related to the above universes, where

$E_{U_1} = \{e_{U_{1,1}} = \text{expensive}, e_{U_{1,2}} = \text{cheap}, e_{U_{1,3}} = \text{wooden}, e_{U_{1,4}} = \text{in green surroundings}\}$
$E_{U_2} = \{e_{U_{2,1}} = \text{expensive}, e_{U_{2,2}} = \text{cheap}, e_{U_{2,3}} = \text{sporty}\}$,
$E_{U_3} = \{e_{U_{3,1}} = \text{expensive}, e_{U_{3,2}} = \text{cheap}, e_{U_{3,3}} = \text{in Kuala Lumpur}, e_{U_{3,4}} = \text{majestic}\}$.

Let $U = \prod_{i=1}^{3} P(U_i)$, $E = \prod_{i=1}^{3} E_{U_i}$ and

$$A^1 = \{e_1 = (e_{U_{1,1}}, e_{U_{2,1}}, e_{U_{3,1}}), e_2 = (e_{U_{1,1}}, e_{U_{2,2}}, e_{U_{3,1}})\},$$
$$A^2 = \{e_1 = (e_{U_{1,1}}, e_{U_{2,1}}, e_{U_{3,1}}), e_3 = (e_{U_{1,2}}, e_{U_{2,3}}, e_{U_{3,1}})\}$$

Suppose that

$$(F^1, A^1) = \{(e_1, (\{h_1, h_2\}, \{c_1, c_2\}, \{v_1\})), (e_2, (\{h_3, h_4\}, \{c_1, c_3\}, \{v_2\}))\},$$
$$(F^2, A^2) = \{(e_1, (\{h_1, h_3\}, \{c_2, c_3\}, \{v_1, v_2\})), (e_3, (\{h_2, h_4\}, \{c_1, c_2\}, \{v_2\}))\},$$
$$(F^3, A^3) = (F^1, A^1) \,\tilde{\cup}\, (F^2, A^2)$$
$$= \{(e_1, (\{h_1, h_2, h_3\}, \{c_1, c_2, c_3\}, \{v_1, v_2\})), (e_2, (\{h_3, h_4\}, \{c_1, c_3\}, \{v_2\})),$$
$$(e_3, (\{h_2, h_4\}, \{c_1, c_2\}, \{v_2\}))\},$$
$$(F^4, A^4) = (F^1, A^1) \,\tilde{\cap}\, (F^2, A^2)$$
$$= \{(e_1, (\{h_1\}, \{c_2\}, \{v_1\})), (e_2, (\{h_3, h_4\}, \{c_1, c_3\}, \{v_2\})),$$
$$(e_3, (\{h_2, h_4\}, \{c_1, c_2\}, \{v_2\}))\},$$

where $A^3 = A^4 = A^1 \cup A^2 = \{e_1 = (e_{U_1,1}, e_{U_2,1}, e_{U_3,1}), e_2 = (e_{U_1,1}, e_{U_2,2}, e_{U_3,1}), e_3 = (e_{U_1,2}, e_{U_2,3}, e_{U_3,1})\}$. Then, we observe that the subfamily $\tau_1 = \{\tilde{\phi}, \tilde{E}, (F^1, A^1), (F^2, A^2), (F^3, A^3), (F^4, A^4)\}$ of SMS(U, E) is a soft multi-topology on (U, E), since it satisfies the necessary three axioms $[O_1]$, $[O_2]$ and $[O_3]$ and $((U, E), \tau_1)$ is a soft multi-topological space. But the subfamily $\tau_2 = \{\tilde{\phi}, \tilde{E}, (F^1, A^1), (F^2, A^2)\}$ of SMS(U, E) is not a soft multi-topology on (U, E) since the union $(F^1, A^1) \,\tilde{\cup}\, (F^2, A^2)$ and the intersection $(F^1, A^1) \,\tilde{\cap}\, (F^2, A^2)$ do not belong to τ_2.

Definition 1.62 Let U be an initial universal set and E be the set of parameters. Let $P(U)$ denote the power set of U & $A \subseteq E$, and then, the pair $\xi = (F, A)$ is called a soft set over U is a parameterised family of subsets of the universe U. For $e \in A$, F $(\underline{e})$ may be considered as a set of e—approximate elements of the soft set (F, A), where $F: A \rightarrow P(U)$.

Definition 1.63 Let U be an initial universe, E be the set of all parameters, and $A \subseteq E$ and $\eta_A(x)$ be a fuzzy set over U for all $x \in E$. Then, a fuzzy soft set (fs set) Γ_A over U is a set defined by a function η_A representing a mapping $\eta_A: E \rightarrow P$ (U) such that $\eta_A(x) = \phi$ if $x \notin A$. Here, η_A is called the fuzzy approximate function of the fs set Γ_A over U and Γ_A can be represented by the set of ordered pairs.

$$\Gamma_A = \{(x, \eta_A(x)): x \in E, \eta_A(x) \in P(U)\}$$

Definition 1.64 Let U be an initial universe, $P(U)$ be the power set of U, E be the set of all parameters, and X be a fuzzy set over E with the membership function μ_X: $E \rightarrow [0, 1]$. Then, the fps set F_X over U is a set defined by the function f_X representing a mapping

$$f_X: E \rightarrow P(U) \text{ such that } f_X(x) = \phi \text{ if } \mu_X(x) = 0$$

Here, f_X is called approximate function of the fps set F_X, and the value f_X is a set called x-element of the fps set for all $x \in E$. Thus, an fps set F_X over U can be represented by the set of pairs.

$$F_X = \{(\mu_X(x)/x, f_X(x)) : x \in E, f_X(x) \in P(U), \mu_X(x) \in [0,1]\}.$$

For example, let $U = \{u_1, u_2, u_3, u_4, u_5\}$ be a universal set and $E = \{x_1, x_2, x_3, x_4\}$ be a set of parameters.

If $X = \{0.2/x_2, 0.5/x_3, 1/x_4\}$ and $f_X(x_2) = \{u_2, u_4\}, f_X(x_3) = \phi, f_X(x_4) = U$;
The fps set F_X is written as $F_X = \{(0.2/x_2, \{u_2, u_4\}), (1/x_4, U)\}$.

Now, the approximate functions of fuzzy parameterised fuzzy soft set (fpfs set) are defined from fuzzy parameter set to the fuzzy subsets of universal set.

Here, we use $\Gamma_X, \Gamma_Y, \Gamma_Z, \ldots$ etc. for fs sets and $\eta_X, \eta_Y, \eta_Z, \ldots$ etc. for their fuzzy approximate functions, respectively.

Definition 1.65 Let U be an initial universe, E be the set of parameters and X be a fuzzy set over E with the membership function $\mu_X: E \to [0,1]$ and $\eta_X(x)$ be a fuzzy set over U for all $x \in E$. Then, a fpfs set Γ_X over U is a set defined by a function $\eta_X(x)$ representing a mapping $\eta_X: E \to P(U)$ such that $\eta_X(x) = \phi$ if $\mu_X(x) = 0$. Here, η_X is called the fuzzy approximate function of the fpfs set Γ_X, and the value $\eta_X(x)$ is a fuzzy set called x-element of the fpfs set for all $x \in E$. Thus, a fpfs set Γ_X over U can be represented by the set of ordered pairs.

$$\Gamma_X = \{(\mu_X(x)/(x)x, \eta_X(x)): x \in E, \eta_X(x) \in P(U), \mu_X(x) \in [0,1]\}.$$

For example, assume that $U = \{u_1, u_2, u_3, u_4, u_5\}$ is a universal set and $E = \{x_1, x_2, x_3, x_4\}$ is a set of parameters. If $X = \{0.2/x_2, 0.5/x_3, 1/x_4\}$ and $\eta_X(x_2) = \{0.5/u_1, 0.3/u_3\}, \eta_X(x_3) = \phi, \eta_X(x_4) = U$, then the fpfs set Γ_X is written as follows:

$$\Gamma_X = \{(0.2/x_1, \{0.5/u_1, 0.3u_3\}, (1/x_4, U))\}.$$

References

1. Chang, C.L.: Fuzzy topological spaces. J. Math. Anal. Appl. **24**, 182–190 (1968)
2. Chakrabarty, K., Gedeon, T., Koczy, J.: Intuitionistic fuzzy rough sets. In: Proceeding of the fourth joint conference of Information Sciences, pp. 211–214. JCTS, Durham, NC (1998)
3. Coker, D.: An introduction to intuitionistic fuzzy topological spaces. Fuzzy Sets Syst. **88**, 81–89 (1997)
4. Jena, S.P., Ghosh, S.K.: Intuitionistic fuzzy rough sets. Notes on Intuitionistic fuzzy sets **8**, 1–18 (2002)
5. Molodtsov, D.: Soft set theory-first results. Comput. Math. Appl. **37**, 19–31 (1999)
6. Nanda, S., Majumder, S.: Fuzzy rough sets. Fuzzy Sets Syst. **45**, 157–160 (1993)
7. Pawlak, Z.: Rough sets. Int. J. Inf. Comput. Sci. **11**, 341–356 (1982)
8. Rizvi, S., Naqvi, H.J., Nadeem, D.: Rough intuitioistic fuzzy sets. In: Proceeding of the 6th joint conference on Information Sciences, pp. 101–104. JCTS, Durham, NC (2002)
9. Samanta, S.K., Mondal, T.K.: Intuitionistic fuzzy rough sets and rough intuitionistic fuzzy sets. J. Fuzzy Math. **9**(3), 561–582 (2001)
10. Zadeh, L.A.: Fuzzy Sets. Inf. Control **8**, 338–353 (1965)

Chapter 2
On Generalised Interval-Valued Intuitionistic Fuzzy Soft Sets

Abstract Molodtsov initiated the concept of fuzzy soft set theory in 1999. Maji et al. introduced the notion of fuzzy soft sets. By introducing the concept of intuitionistic fuzzy sets into the theory of soft sets, Maji et al. proposed the concept of intuitionistic fuzzy soft set theory. The notion of the interval-valued intuitionistic fuzzy sets was first introduced by Atanassov and Gargov. It is characterised by an interval-valued membership degree and an interval-valued non-membership degree. In 2010, Y. Jiang et al. introduced the concept of interval-valued intuitionistic fuzzy soft sets. In this chapter, the concept of generalised interval-valued intuitionistic fuzzy soft sets is introduced. The basic properties of these sets are presented. Also, an application of generalised interval-valued intuitionistic fuzzy soft sets in decision-making with respect to interval of degree of preference is investigated.

Keywords Soft sets · Fuzzy soft sets · Interval-valued fuzzy sets · Intuitionistic fuzzy sets · Intuitionistic fuzzy soft sets · Generalised intuitionistic fuzzy soft sets · Interval-valued intuitionistic fuzzy sets · Interval-valued intuitionistic fuzzy soft sets · Generalised interval-valued intuitionistic fuzzy soft sets

In 1999, Molodtsov [9] initiated the concept of fuzzy soft set theory, which is completely a new approach for modelling vagueness and uncertainties. Soft set theory has a rich potential for application in solving various decision-making problems. Maji et al. [6] introduced the concept of fuzzy soft set theory. As a generalisation of fuzzy soft set theory, intuitionistic fuzzy soft set theory [7] makes description of the objective more realistic, more practical, and accurate in some cases, making it more promising. After the introduction of fuzzy set [10], Atanassov [1], introduced intuitionistic fuzzy set as a generalisation fuzzy set. Gorzalczany [4], introduced the interval-valued fuzzy set in 1987. In 2010, Majumder and Samanta [8], introduced generalised fuzzy soft set. Also in 2010, Dinda, Bera and Samanta [3], introduced generalised fuzzy soft set. Atanassov and Gargov [2] introduced the concept of interval-valued intuitionistic fuzzy set theory. In 2010, Jiang et al. [5] introduced the concept of interval-valued intuitionistic fuzzy soft sets which is a combination of an interval-valued intuitionistic fuzzy set theory and a soft set theory. In this chapter, the concept of generalised interval-valued intuitionistic fuzzy soft sets

© Springer India 2015   23
A. Mukherjee, *Generalized Rough Sets*, Studies in Fuzziness
and Soft Computing 324, DOI 10.1007/978-81-322-2458-7_2

together with their basic properties is introduced. Also, an application of generalised interval-valued intuitionistic fuzzy soft sets in decision-making is presented.

Throughout the text, unless otherwise stated explicitly, U be the set of universe and E be the set of parameters, and we take $A, B, C \subseteq E$ and α, β, δ are fuzzy subsets of A, B, C, respectively.

Definition 2.1 Let U be an initial universe and E be a set of parameters. Let IVIFS^U be the set of all interval-valued intuitionistic fuzzy soft sets on U and $A \subseteq E$. Let F be a mapping given by $F: A \to \mathrm{IVIFS}^U$ and α be a mapping given by $\alpha: A \to \mathrm{Int}([0, 1])$. Let F_α be a mapping given by $F: A \to \mathrm{IVIFS}^U \times \mathrm{Int}([0, 1])$ and defined by

$$F_\alpha(e) = (F(e), \alpha(e))$$
$$= \left(\langle x, \mu_{F(e)}(x), \gamma_{F(e)}(x) \rangle, \alpha(e) \right)$$

where $e \in A$ and $x \in U$ where $\alpha(e) = [[\alpha(e)\downarrow, \alpha(e)\uparrow]]$.

Here, $\mu_{F(e)}(x)$ is the interval-valued fuzzy membership degree that object x holds on parameter ε and $\gamma_{F(e)}(x)$ is the interval-valued fuzzy membership degree that object x does not hold on parameter ε. For each parameter e, $\alpha(e)$ will be termed as the interval of degree of preference. The pair (F_α, A) is called a generalised interval-valued intuitionistic fuzzy soft set over (U, E).

Example 2.2 Let $U = \{h_1, h_2, h_3, h_4, h_5\}$ be the set of five houses under the consideration of a decision-maker to purchase. Let $A \subseteq E$ and $A = \{e_1(\text{beautiful}), e_2(\text{expensive}), e_3(\text{wooden}), e_4(\text{in good repair}), e_5(\text{in green surroundings})\}$. Let $\alpha: A \to \mathrm{Int}([0, 1])$ be defined by

$$\alpha(e_1) = [0.6, 0.8], \quad \alpha(e_2) = [0.5, 0.7], \quad \alpha(e_3) = [0.4, 0.5],$$
$$\alpha(e_4) = [0.3, 0.45], \quad \alpha(e_5) = [0.2, 0.5].$$

Now, we define F_α as follows:

$F_\alpha(e_1) = (\{\langle h_1, [0.5, 0.7], [0.1, 0.2]\rangle, \langle h_2, [0.7, 0.8], [0.05, 0.1]\rangle, \langle h_3, [0.6, 0.7], [0.2, 0.24]\rangle,$
$\quad \langle h_4, [0.3, 0.4], [0.4, 0.5]\rangle, \langle h_5, [0.01, 0.05], [0.07, 0.09]\rangle\}, [0.6, 0.8])$

$F_\alpha(e_2) = (\{\langle h_1, [0.7, 0.8], [0.1, 0.2]\rangle, h_2, [0.5, 0.6], [0.2, 0.3]\rangle, \langle h_3, [0.4, 0.6], [0.3, 0.37]\rangle,$
$\quad \langle h_4, [0.1, 0.3], [0.4, 0.5]\rangle, \langle h_5, [0.55, 0.7], [0.25, 0.29]\rangle\}, [0.5, 0.7])$

$F_\alpha(e_3) = (\{\langle h_1, [0.3, 0.4], [0.4, 0.5]\rangle, \langle h_2, [0.65, 0.75], [0.01, 0.23]\rangle, \langle h_3, [0.55, 0.7], [0.2, 0.25]\rangle,$
$\quad \langle h_4, [0.6, 0.8], [0.1, 0.2]\rangle, \langle h_5, [0.3, 0.6], [0.1, 0.2]\rangle\}, [0.4, 0.5])$

$F_\alpha(e_4) = (\{\langle h_1, [0.1, 0.3], [0.5, 0.6]\rangle, \langle h_2, [0.25, 0.75], [0.05, 0.2]\rangle, \langle h_3, [0.6, 0.7], [0.1, 0.2]\rangle,$
$\quad \langle h_4, [0.1, 0.4], [0.2, 0.5]\rangle, \langle h_5, [0.4, 0.5], [0.2, 0.35]\rangle\}, [0.3, 0.45])$

$F_\alpha(e_5) = (\{\langle h_1, [0.2, 0.4], [0.3, 0.5]\rangle, \langle h_2, [0.3, 0.4], [0.35, 0.55]\rangle, \langle h_3, [0.5, 0.6], [0.05, 0.15]\rangle,$
$\quad \langle h_4, [0.6, 0.7], [0.1, 0.2]\rangle, \langle h_5, [0.1, 0.5], [0.2, 0.3]\rangle\}, [0.2, 0.5])$

Here, (F_α, A) is a generalised interval-valued intuitionistic fuzzy soft set over (U, E).

Definition 2.3 Let (F_α, A) and (G_β, B) be two generalised interval-valued intuitionistic fuzzy soft sets over (U, E). Then, (F_α, A) is called a generalised interval-valued intuitionistic fuzzy soft subset of (G_β, B), denoted by $(F_\alpha, A) \subseteq (G_\beta, B)$ if

(a) $A \subseteq B$
(b) $\forall e \in A, \alpha(e) \subseteq \beta(e)$
(c) $\forall e \in A, F(e)$ is an interval-valued intuitionistic fuzzy subset of $G(e)$.

Definition 2.4 Let (F_α, A) and (G_β, B) be two generalised interval-valued intuitionistic fuzzy soft sets over (U, E). Then, the intersection of (F_α, A) and (G_β, B) is a generalised interval-valued intuitionistic fuzzy soft sets over (U, E), denoted by $(F_\alpha, A) \cap (G_\beta, B)$, and is defined by $(F_\alpha, A) \cap (G_\beta, B) = (H_\delta, A \cap B)$, where $H_\delta: A \cap B \rightarrow \text{IVIFS}^U \times \text{Int}([0, 1])$ is a mapping such that $\forall e \in A \cap B$ and $x \in U$,

$$H_\delta(e) = \left(\left\langle x, \mu_{H(e)}(x), \gamma_{H(e)}(x) \right\rangle, \delta(e) \right),$$

where

$$\mu_{H(e)}(x) = \left[\inf\left(\underline{\mu}_{F(e)}(x), \underline{\mu}_{G(e)}(x) \right), \inf\left(\overline{\mu}_{F(e)}(x), \overline{\mu}_{G(e)}(x) \right) \right],$$

$$\gamma_{H(e)}(x) = \left[\sup\left(\underline{\gamma}_{F(e)}(x), \underline{\gamma}_{G(e)}(x) \right), \sup\left(\overline{\gamma}_{F(e)}(x), \overline{\gamma}_{G(e)}(x) \right) \right] \text{ and}$$

$$\delta(e) = \alpha(e) * \beta(e)[\alpha(e)\downarrow \cdot \beta(e)\downarrow, \alpha(e)\uparrow \cdot \beta(e)\uparrow].$$

Definition 2.5 Let (F_α, A) and (G_β, B) be two generalised interval-valued intuitionistic fuzzy soft sets over (U, E). Then, the union of (F_α, A) and (G_β, B) is a generalised interval-valued intuitionistic fuzzy soft sets over (U, E), denoted by $(F_\alpha, A) \cup (G_\beta, B)$, and is defined by

$$(F_\alpha, A) \cup (G_\beta, B) = (H_\delta, A \cup B)$$

where $H_\delta: A \cup B \rightarrow \text{IVIFS}^U \times \text{Int}([0, 1])$ is a mapping such that $\forall e \in A \cup B$ and $x \in U$,

$$H_\delta(e) = \left(\left\langle x, \mu_{F(e)}(x), \gamma_{F(e)}(x) \right\rangle, \alpha(e) \right) \quad \text{if } e \in A - B$$

$$= \left(\left\langle x, \mu_{G(e)}(x), \gamma_{G(e)}(x) \right\rangle, \beta(e) \right) \quad \text{if } e \in B - A$$

$$= \left(\left\langle x, \mu_{H(e)}(x), \gamma_{H(e)}(x) \right\rangle, \delta(e) \right) \quad \text{if } e \in A \cap B,$$

where

$$\mu_{H(e)}(x) = \left[\inf\left(\underline{\mu}_{F(e)}(x), \underline{\mu}_{G(e)}(x) \right), \inf\left(\overline{\mu}_{F(e)}(x), \overline{\mu}_{G(e)}(x) \right) \right],$$

$$\gamma_{H(e)}(x) = \left[\sup\left(\underline{\gamma}_{F(e)}(x), \underline{\gamma}_{G(e)}(x) \right), \sup\left(\overline{\gamma}_{F(e)}(x), \overline{\gamma}_{G(e)}(x) \right) \right] \text{ and}$$

$$\delta(e) = \alpha(e)\Delta\beta(e) = [\alpha(e){\downarrow} + \beta(e){\downarrow} - \alpha(e){\downarrow} \cdot \beta(e){\downarrow}, \alpha(e){\uparrow} + \beta(e){\uparrow} - \alpha(e){\uparrow} \cdot \beta(e){\uparrow}]$$

Proposition 2.6 *Let (F_α, A), (G_β, B), and (H_δ, C) be three generalised interval-valued intuitionistic fuzzy soft sets over (U, E). Then*

(a) $(F_\alpha, A) \cup (G_\beta, B) = (G_\beta, B) \cup (F_\alpha, A)$
(b) $(F_\alpha, A) \cap (G_\beta, B) = (G_\beta, B) \cap (F_\alpha, A)$
(c) $(F_\alpha, A) \cup ((G_\beta, B) \cup (H_\delta, C)) = ((F_\alpha, A) \cup (G_\beta, B)) \cup (H_\delta, C)$
(d) $(F_\alpha, A) \cap ((G_\beta, B) \cap (H_\delta, C)) = ((F_\alpha, A) \cap (G_\beta, B)) \cap (H_\delta, C)$
(e) $(F_\alpha, A) \cup ((G_\beta, B) \cap (H_\delta, C)) = ((F_\alpha, A) \cup (G_\beta, B)) \cap ((F_\alpha, A) \cup ((H_\delta, C))$
(f) $(F_\alpha, A) \cap ((G_\beta, B) \cup (H_\delta, C)) = ((F_\alpha, A) \cap (G_\beta, B)) \cup ((F_\alpha, A) \cap ((H_\delta, C)).$

Definition 2.7 Let (F_α, A) be a generalised interval-valued intuitionistic fuzzy soft set over (U, E). Then, the complement of (F_α, A) is a generalised interval-valued intuitionistic fuzzy soft sets over (U, E), denoted by $(F_\alpha, A)^c$, and is defined by

$$(F_\alpha, A)^c = (F_\alpha^c, A^c)$$

where $F_\alpha^c: A^c \rightarrow \text{IVIFS}^U \times \text{Int}([0, 1])$ is a function defined by

$$F_\alpha^c(\sim e) = \left(\left\langle x, \gamma_{F(e)}(x), \mu_{F(e)}(x) \right\rangle, \alpha^c(e) \right)$$

where $\sim e \in A^c$ and $x \in U$ and $\alpha^c: A^c \rightarrow \text{Int}([0, 1])$ is defined as

$$\alpha^c(\sim e) = [1\alpha(e){\uparrow} -, 1 - \alpha(e){\downarrow}] \quad \text{if } \alpha(e) = [\alpha(e){\downarrow}, \alpha(e){\uparrow}] \in \text{Int}[0, 1], \ e \in A.$$

Proposition 2.8: *Let (F_α, A) be a generalised interval-valued intuitionistic fuzzy soft set over (U, E). Then, $[(F_\alpha, A)^c]^c = (F_\alpha, A)$.*

2.1 An Application of Generalised Interval-Valued Intuitionistic Fuzzy Soft Sets in Decision-Making

Let us consider the Example 2.2. Let $E = \{e_1(\text{beautiful}), e_2(\text{expensive}), e_3(\text{wooden}), e_4(\text{in good repair}), e_5(\text{in green surroundings}), e_6(\text{cheap}), e_7(\text{modern})\}$.

The problem is that out of the available houses in U, we have to select that house which qualifies with all or maximum number of parameters of the parameter set A. Now, we introduce the following two operations:

- *For interval-valued fuzzy membership degree*:

For $e \in A, x \in U$ and $\alpha(e) = [\alpha(e)\downarrow, \alpha(e)\uparrow]$

$$\mu'_{F(e)}(x) = \left[\underline{\mu}_{F(e)}(x) + \alpha(e)\downarrow - \underline{\mu}_{F(e)}(x) \cdot \alpha(e)\downarrow, \overline{\mu}_{F(e)}(x) + \alpha(e)\downarrow - \overline{\mu}_{F(e)}(x) \cdot \alpha(e)\downarrow\right]$$

- *For interval-valued fuzzy non-membership degree*:

For $e \in A, x \in U$ and $\alpha(e) = [\alpha(e)\downarrow, \alpha(e)\uparrow]$,

$$\gamma'_{F(e)} = \left[\underline{\gamma}_{F(e)}(x) \cdot \alpha(e)\uparrow, \underline{\gamma}_{F(e)}(x) \cdot \alpha(e)\uparrow\right] \quad \text{if} \quad \overline{\mu}_{F(e)}(x) + \alpha(e)\downarrow$$
$$- \overline{\mu}_{F(e)}(x) \cdot \alpha(e)\downarrow + \overline{\gamma}_{F(e)}(x) \cdot \alpha(e)\uparrow \leq 1$$
$$= [0, 0], \text{otherwise}$$

Actually, we have taken these two operations to ascend the interval-valued fuzzy membership degree and to descend the interval-valued fuzzy non-membership degree on the basis of the interval of degree of preference. Then, the generalised interval-valued intuitionistic fuzzy soft set (F_α, A) is reduced to an interval-valued intuitionistic fuzzy soft set (F', A) which is given as follows:

$F'(e_1) = \{\langle h_1, [0.8, 0.88], [0, 0]\rangle, \langle h_2, [0.88, 0.92], [0.04, 0.08]\rangle,$
$\qquad \langle h_3, [0.84, 0.88], [0, 0]\rangle, \langle h_4, [0.72, 0.76], [0, 0]\rangle,$
$\qquad \langle h_5, [0.604, 0.62], [0.056, 0.072]\rangle\}$

$F'(e_2) = \{\langle h_1, [0.85, 0.90], [0, 0]\rangle, \langle h_2, [0.75, 0.80], [0, 0]\rangle, \langle h_3, [0.70, 0.80], [0, 0]\rangle,$
$\qquad \langle h_4, [0.55, 0.65], [0.28, 0.35]\rangle, \langle h_5, [0.775, 0.85], [0.0]\rangle\}$

$F'(e_3) = \{\langle h_1, [0.58, 0.64], [0.20, 0.25]\rangle, \langle h_2, [0.79, 0.85], [0.005, 0.115]\rangle,$
$\qquad \langle h_3, [0.73, 0.82], [0.10, 0.125]\rangle, \langle h_4, [0.76, 0.88], [0.05, 0.10]\rangle,$
$\qquad \langle h_5, [0.58, 0.76], [0.05, 0.10]\rangle\}$

$F'(e_4) = \{\langle h_1, [0.37, 0.51], [0.225, 0.270]\rangle, \langle h_2, [0.475, 0.825], [0.0225, 0.09]\rangle,$
$\qquad \langle h_3, [0.72, 0.79], [0.045, 0.09]\rangle, \langle h_4, [0.37, 0.58], [0.09, 0.225]\rangle,$
$\qquad \langle h_5, [0.58, 0.65], [0.09, 0.1575]\rangle\}$

$F'(e_5) = \{\langle h_1, [0.36, 0.52], [0.15, 0.25]\rangle, \langle h_2, [0.44, 0.52], [0.175, 0.275]\rangle,$
$\qquad \langle h_3, [0.60, 0.68], [0.025, 0.075]\rangle, \langle h_4, [0.68, 0.76], [0.05, 0.10]\rangle,$
$\qquad \langle h_5, [0.28, 0.60], [0.10, 0.15]\rangle\}$

Now to reduce the above interval-valued intuitionistic fuzzy soft set (F', A) into an intuitionistic fuzzy soft set (F'', A), we apply the following two operations:

- *For membership function*:

$$\mu''_{F(e)}(x) = \left(\inf \mu'_{F(e)}(x), \sup \mu'_{F(e)}(x)\right)/2, \quad \text{for } e \in A \ \& \ x \in U$$

- *For non-membership function*:

$$\gamma''_{F(e)}(x) = \left(\inf \gamma'_{F(e)}(x), \sup \gamma'_{F(e)}(x)\right)/2, \quad \text{for } e \in A \ \& \ x \in U$$

Then, the reduced intuitionistic fuzzy soft set (F'', A) is given as follows:

$$F''(e_1) = \{\langle h_1, 0.84, 0\rangle, \langle h_2, 0.90, 0.06\rangle, \langle h_3, 0.86, 0\rangle,$$
$$\langle h_4, 0.74, 0\rangle, \langle h_5, 0.612, 0.064\rangle\}$$
$$F''(e_2) = \{\langle h_1, 0.875, 0\rangle, \langle h_2, 0.775, 0\rangle, \langle h_3, 0.75, 0\rangle,$$
$$\langle h_4, 0.60, 0.315\rangle, \langle h_5, 0.8125, 0\rangle\}$$
$$F''(e_3) = \{\langle h_1, 0.61, 0.225\rangle, \langle h_2, 0.82, 0.06\rangle, \langle h_3, 0.775, 0.1125\rangle,$$
$$\langle h_4, 0.82, 0.075\rangle, \langle h_5, 0.67, 0.075\rangle\}$$
$$F''(e_4) = \{\langle h_1, 0.44, 0.2475\rangle, \langle h_2, 0.65, 0.05625\rangle, \langle h_3, 0.755, 0.0675\rangle,$$
$$\langle h_4, 0.475, 0.1575\rangle, \langle h_5, 0.615, 0.12375\rangle\}$$
$$F''(e_5) = \{\langle h_1, 0.44, 0.20\rangle, \langle h_2, 0.48, 0225\rangle, \langle h_3, 0.64, 0.05\rangle,$$
$$\langle h_4, 0.72, 0.075\rangle, \langle h_5, 0.44, 0.125\rangle\}.$$

Algorithm

1. Input the set A of choice of parameters.
2. Consider the reduced intuitionistic fuzzy soft set in tabular form.
3. Compute the comparison table for both membership and non-membership function.
4. Compute the membership and non-membership score of each object.
5. Compute the final score.
6. If the maximum score occurs in the ith row, then the house h_i will be purchased.

Let us use the algorithm to solve the problem (Tables 2.1, 2.2, 2.3, 2.4, 2.5 and 2.6).

Table 2.1 Tabular representation of intuitionistic fuzzy soft set (F'', A)

	e_1	e_2	e_3	e_4	e_5
h_1	(0.84, 0)	(0.875, 0)	(0.61, 0.225)	(0.44, 0.2475)	(0.44, 0.20)
h_2	(0.90, 0.06)	(0.775, 0)	(0.82, 0.06)	(0.65, 0.05625)	(0.48, 0.225)
h_3	(0.86, 0)	(0.75, 0)	(0.775, 0.1125)	(0.755, 0.0675)	(0.64, 0.05)
h_4	(0.74, 0)	(0.60, 0.315)	(0.82, 0.075)	(0.475, 0.1575)	(0.72, 0.075)
h_5	(0.612, 0.064)	(0.8125, 0)	(0.67, 0.075)	(0.615, 0.12375)	(0.44, 0.125)

Table 2.2 Comparison table for membership function

	h_1	h_2	h_3	h_4	h_5
h_1	5	1	1	2	3
h_2	4	5	3	4	4
h_3	4	2	5	3	4
h_4	3	2	2	5	3
h_5	3	1	1	2	5

Table 2.3 Comparison table for non-membership function

	$h\,h_1$	$h\,h_2\,h_2$	$h\,h_3$	$h\,h_4\,h_4$	$h\,h_5\,h_5$
h_1	5	3	5	4	4
h_2	3	5	3	2	2
h_3	2	3	5	2	2
h_4	2	3	4	5	3
h_5	2	4	4	3	5

Table 2.4 Membership score table

	Row sum	Column sum	Membership score
h_1	12	19	−7
h_2	20	11	9
h_3	18	12	6
h_4	15	16	−1
h_5	12	19	−7

Table 2.5 Non-membership score table

	Row sum	Column sum	Non-membership score
h_1	21	14	7
h_2	15	18	−3
h_3	14	21	−7
h_4	17	16	1
h_5	18	16	2

Table 2.6 Final score table

	Membership score	Non-membership score	Final score
h_1	−7	7	−14
h_2	9	−3	12
h_3	6	−7	13
h_4	−1	1	−2
h_5	−7	2	−9

2.2 Conclusion

As the maximum score is 13, so the decision-maker will purchase 'house h_3'.

References

1. Atanassov, K.: Intuitionistic fuzzy sets. Fuzzy Sets Syst. **20**, 87–96 (1986)
2. Atanassov, K., Gargov, G.: Interval-valued intuitionistic fuzzy sets. Fuzzy Sets Syst. **31**, 343–349 (1989)
3. Dinda, B., Bera, T., Samanta, T.K.: Generalised Intuitionistic fuzzy soft sets and its application in decision making. arXiv:1010.2468vI[math.GM], 12 Oct 2010 [online]
4. Gorzalczany, M.: A method of inference in approximate reasoning based on interval-valued fuzzy sets. Fuzzy Sets Syst. **21**, 1–17 (1987)
5. Jiang, Y., Tang, Y., Chen, Q., Liu, H., Tang, J.: Interval-valued intuitionistic fuzzy soft sets and their properties. Comput. Math. Appl. **60**, 906–918 (2010)
6. Maji, P.K., Roy, A.R., Biswas, R.: Fuzzy soft sets. J. Fuzzy Math. **9**(3), 589–602 (2001)
7. Maji, P.K., Roy, A.R., Biswas, R.: On intuitionistic fuzzy soft sets. J. Fuzzy Math. **12**(3), 669–683 (2004)
8. Majumder, P., Samanta, S.K.: Generalised fuzzy soft sets. Comput. Math. Appl. **59**(4), 1425–1432 (2010)
9. Molodtsov, D.: Soft set theory-first results. Comput. Math. Appl. **37**, 19–31 (1999)
10. Zadeh, L.A.: Fuzzy sets. Inf. Control **8**, 338–353 (1965)

Chapter 3
Soft Rough Intuitionistic Fuzzy Sets

Abstract Theories of fuzzy sets and rough sets are powerful mathematical tools for modelling various types of uncertainty. Molodtsov (Comput Math Appl 37:19–31, 1999 [6]) initiated a novel concept called soft sets, a new mathematical tool for dealing with uncertainties. It has been found that fuzzy sets, rough sets, and soft sets are closely related concepts (Aktas and Cagman in Inf Sci 1(77):2726–2735, 2007 [1]). Research works on soft sets are very active and progressing rapidly in these years. In 2001, Maji et al. (J Fuzzy Math 9(3):589–602, 2001 [5]) proposed the idea of intuitionistic fuzzy soft set theory and established some results on them. Based on an equivalence relation on the universe of discourse, Dubois and Prade (Int J Gen Syst 17:191–209, 1990 [3]) introduced the lower and upper approximation of fuzzy sets in a Pawlak approximation space and obtained a new notion called rough fuzzy sets. Feng et al. (Soft Compt 14:899–911, 2009 [4]) introduced lower and upper soft rough approximation of fuzzy sets in a soft approximation space and obtained a new hybrid model called soft rough fuzzy sets which is the extension of Dubois and Prade's rough fuzzy sets. The aim of this chapter is to consider lower and upper soft rough intuitionistic fuzzy approximation of intuitionistic fuzzy sets in intuitionistic fuzzy soft approximation space (IF soft approximation space) and obtain a new hybrid model called soft rough intuitionistic fuzzy sets which can be seen as extension of both the previous work by Dubois and Prade and Feng et al.

Keywords Fuzzy set · Soft set · Rough set · Soft rough fuzzy set · Intuitionistic fuzzy soft set

Theories of fuzzy sets and rough sets are powerful mathematical tools for modelling various types of uncertainty. Molodtsov [6] initiated a novel concept called soft sets a new mathematical tool for dealing with uncertainties. It has been found that fuzzy sets, rough sets and soft sets are closely related concepts [1]. Research works on soft sets are very active and progressing rapidly in these years. Based on an equivalence relation on the universe of discourse, Dubois and Prade [3] introduced the lower and upper approximation of fuzzy sets in a Pawlak [7] approximation space and obtained a new notion called rough fuzzy sets. Feng et al. [4] introduced lower and upper soft rough approximation of fuzzy sets in a soft approximation

© Springer India 2015

A. Mukherjee, *Generalized Rough Sets*, Studies in Fuzziness
and Soft Computing 324, DOI 10.1007/978-81-322-2458-7_3

space and obtained a new hybrid model called soft fuzzy sets which is the extension
of Dubois and Prade's rough fuzzy sets. Many fields deal daily with the uncertain
data that may not be successfully modelled by the classical mathematics. There are
some mathematical tools for dealing with uncertainties, two of them are fuzzy set
theory, developed by Zadeh [8], and soft set theory, introduced by Molodtsov [6]
that are related to our work. On the other hand, a fuzzy set allows a membership
value other than 0 and 1. A rough set uses there membership functions, a reference
set, and its lower and upper approximation in an approximation space. There are
extensive studies on the relationships between rough sets and fuzzy sets. Many
proposals have been made for the combination of rough set and fuzzy set. The result
of these studies leads to the introduction of the notion of rough fuzzy sets and fuzzy
rough sets. Dubois and Prade investigated the problem of combining fuzzy sets with
rough sets. In general, a rough fuzzy set is the approximation of a fuzzy set in a
crisp approximation space, whereas a fuzzy rough set is the approximation of a
crisp set or a fuzzy set in a fuzzy approximation space. Based on a Pawlak
approximation space, the approximation of a soft set was proposed to obtain a
hybrid model called rough soft sets [4]. Alternatively a soft set instead of an
equivalence relation was used to generalise the universe. This lead to a deviation of
Pawlak approximation space called a soft approximation space in which soft rough
approximation and soft rough sets were introduced [4]. Further, they considered
approximation of a fuzzy set in a soft approximation space and initiated a concept
called soft rough fuzzy sets which was the extension of Dubois and Prade's rough
fuzzy sets. Atanassov [2], introduced the concept of intuitionistic fuzzy set in 1986.

The aim of this chapter was to introduce lower and upper soft approximation of
intuitionistic fuzzy sets in a soft approximation space and obtain a new hybrid
model called soft rough intuitionistic fuzzy sets.

Definition 3.1 A soft set $\xi = (F, A)$ over the universe U is called, a full soft set if
$\underset{e \in A}{\cup} F(e) = U$.

Definition 3.2 A full soft set $\xi = (F, A)$ over U called a covering soft set if $F(e) \neq \Phi, \forall e \in A$.

Definition 3.3 Let $\xi = (F, A)$ be a full soft set over U and $S = (U, \xi)$ be a soft
approximation space. For an IF set, $\alpha \in P(U)$ ($P(U)$ is the family of all intuitionistic
fuzzy sets in U), the lower and upper soft rough approximations of α with respect to
S are defined by $Sap_S(\alpha)\downarrow$ and $Sap_S(\alpha)\uparrow$, respectively, which are IF sets in U
given by

$$Sap_S(\alpha) \downarrow = \wedge\{\alpha(y) : \exists e \in A/\{x,y\} \subseteq F(e)\}, \quad \text{and}$$
$$Sap_S(\alpha) \uparrow = \vee\{\alpha(y) : \exists e \in A/\{x,y\} \subseteq F(e)\} \quad \text{for all} \quad x \in U.$$

The operators $Sap_S\downarrow$ and $Sap_S\uparrow$ are called the lower and upper soft rough
approximation operators on IF sets. If $IFSap_S\downarrow = IFSap_S\uparrow$, α is said to be soft

definable; otherwise, α is called a soft rough intuitionistic fuzzy set. Here, $\alpha(y) = \{(y, \mu_A(y), \nu_A(y) : y \in X)\}$. More over, $\boldsymbol{P} \text{ os}_S(x) = \text{Sap}_S(x)\downarrow$, $\boldsymbol{N} \text{ eg}_S(x) = U - \text{Sap}_S(x)\uparrow$, and $\boldsymbol{B} \text{ nd}_S(x) = \text{Sap}_S(x)\uparrow - \text{Sap}_S(x)\downarrow$ are called the soft positive, soft negative, and soft boundary regions of X, respectively.

By definition, we immediately have that $X \subseteq U$ is a soft definable set if $\boldsymbol{B} \text{ nd}_S(x) = \Phi$. Also it is clear that $\text{Sap}_S(x)\downarrow \subseteq X$ and $\text{Sap}_S(x)\downarrow \subseteq \text{Sap}_S(x)\uparrow$ for all $X \subseteq U$.

Theorem 3.4 *Let $\xi = (F, A)$ be a soft set over U, $S = \{U, \xi\}$ be a soft approximation space, and $\alpha, \beta \in P(U)$, then we have the following:*

(a) $\text{Sap}_S(\alpha)\downarrow \subseteq \alpha \subseteq \text{Sap}_S(\alpha)\uparrow$.
(b) $\text{Sap}_S(\Phi)\downarrow = \text{Sap}_S(\Phi)\uparrow = \Phi$.
(c) $\text{Sap}_S(U)\downarrow = \text{Sap}_S(U)\uparrow = U$.
(d) $(\text{Sap}_S(\alpha)\uparrow)^c = (\text{Sap}_S(\alpha^c))\downarrow$, if $\alpha = \{(x, \mu_A(x), \nu_A(x) : x \in U)\}$ then $\alpha^c = \{(x, \nu_A(x), \mu_A(x) : x \in U)\}$.
(e) $(\text{Sap}_S(\alpha)\downarrow)^c = (\text{Sap}_S(\alpha^c))\uparrow$.
(f) $(\text{Sap}_S(\alpha \cap \beta)\downarrow) = \text{Sap}_S(\alpha)\downarrow \cap \text{Sap}_S(\beta)\downarrow$.
(g) $(\text{Sap}_S(\alpha \cup \beta)\downarrow) \supseteq \text{Sap}_S(\alpha)\downarrow \cup \text{Sap}_S(\beta)\downarrow$.
(h) $(\text{Sap}_S(\alpha \cup \beta)\uparrow) = \text{Sap}_S(\alpha)\uparrow \cup \text{Sap}_S(\beta)\uparrow$.
(i) $(\text{Sap}_S(\alpha \cap \beta)\uparrow) \subseteq \text{Sap}_S(\alpha)\uparrow \cap \text{Sap}_S(\beta)\uparrow$.
(j) $\alpha \subseteq \beta \Rightarrow \text{Sap}_S(\alpha)\downarrow \subseteq \text{Sap}_S(\beta)\downarrow$.
(k) $\alpha \subseteq \beta \Rightarrow \text{Sap}_S(\alpha)\uparrow \subseteq \text{Sap}_S(\beta)\uparrow$.

Proof

(a) Let $\alpha \in P(U)$ and $x \in U$. Since $\xi = (F, A)$ is a soft set over U, there exists some $e_0 \in A$ such that $x \in F(e_0)$. Now by the definitions, we have $\text{Sap}_S(\alpha)\downarrow(x) = \wedge\{\alpha(y) : \exists e \in A/\{x, y\} \subseteq F(e)\}$, and $\text{Sap}_S(\alpha)\uparrow(x) = \vee\{\alpha(y) : \exists e \in A/\{x, y\} \subseteq F(e)\}$ for all $x \in U$. Hence, it follows that $\text{Sap}_S(\alpha)\downarrow(x) \leq \alpha(x) \leq \text{Sap}_S(\alpha)\uparrow(x)$. This shows that $\text{Sap}_S(\alpha)\downarrow \subseteq \alpha \subseteq \text{Sap}_S(\alpha)\uparrow$.
(b) This proof is straightforward.
(c) This proof is also straightforward.
(d) Let $\alpha \in P(U)$ and $x \in U$ and let $\alpha = \{(x, \mu_\alpha(x), \nu_\alpha(x) : x \in U)\}$, $\beta = \{(x, \mu_\beta(x), \nu_\beta(x) : x \in U)\}$. Take $N(x) = \{y : \exists e \in A/\{x, y\} \subseteq F(e)\}$ for all $x \in U$. Now, $\text{Sap}_S(\alpha)\uparrow(x) = \vee\{\alpha(y) : y \in N(x)\}$. Hence, $\alpha(y) \leq \text{Sap}_S(\alpha)\uparrow(x)$ for all $y \in N(x)$. Now it follows that $(\text{Sap}_S(\alpha)\uparrow)^c = 1 - (\text{Sap}_S(\alpha)\uparrow(x)) \leq 1 - \alpha(y) = \alpha^c(y)$ for all $y \in N(x)$. Thus we have

$$(\text{Sap}_S(\alpha)\uparrow)^c(x) \leq \wedge\{\alpha(y) : y \in N(x)\} = (\text{Sap}_S(\alpha^c))\downarrow(x).$$

Thus,

$$(\text{Sap}_S(\alpha)\uparrow)^c \subseteq (\text{Sap}_S(\alpha^c))\downarrow. \qquad (i)$$

By using the similar method, we can prove that $(\mathrm{Sap}_S(\beta)\!\downarrow)^c \supseteq (\mathrm{Sap}_S(\beta^c))\!\uparrow$ for all intuitionistic fuzzy set $\beta \in P(U)$. Now taking $\beta = \alpha^c$, we get $(\mathrm{Sap}_S(\alpha^c)\!\downarrow)^c \supseteq (\mathrm{Sap}_S(\alpha))\!\uparrow$, that is

$$(\mathrm{Sap}_S(\alpha^c)\!\downarrow) \subseteq (\mathrm{Sap}_S(\alpha)\!\uparrow)^c. \qquad \text{(ii)}$$

From (i) and (ii), $(\mathrm{Sap}_S(\alpha)\!\uparrow)^c = (\mathrm{Sap}_S(\alpha^c))\!\downarrow$.

(e) This is similar to the proof of (d).

(f) Let $\alpha, \beta \in P(U)$, $x \in U$ and let $\alpha = \{(x, \mu_\alpha(x), v_\alpha(x) : x \in U)\}$, $\beta = \{(x, \mu_\beta(x), v_\beta(x) : x \in U)\}$. Put $N(x) = \{y : \exists e \in A/\{x, y\} \subseteq F(e)\}$. Now, $\alpha \cap \beta = \{(x, \mu_\alpha(x) \cap \mu_\beta(x), v_\alpha(x) \cup v_\beta(x) : x \in U)\}$. So $\mathrm{Sap}_S(\alpha \cap \beta)\!\downarrow(x) = \wedge\{\alpha(y) \cap \beta(y) : y \in N(x)\}$, Hence, $\mathrm{Sap}_S(\alpha \cap \beta)\!\downarrow \subseteq \alpha(y) \cap \beta(y) \subseteq \alpha(y)$ for all $y \in N(x)$. Now, $\mathrm{Sap}_S(\alpha)\!\downarrow(x) = \wedge\{\alpha(y) : y \in N(x)\}$. This implies $\mathrm{Sap}_S(\alpha \cap \beta)\!\downarrow(x) \leq \mathrm{Sap}_S(\alpha)\!\downarrow(x)$. Similarly, $\mathrm{Sap}_S(\alpha \cap \beta)\!\downarrow(x) \leq \mathrm{Sap}_S(\beta)\!\downarrow(x)$. Therefore, $\mathrm{Sap}_S(\alpha \cap \beta)\!\downarrow(x) \leq \mathrm{Sap}_S(\alpha)\!\downarrow(x) \cap \mathrm{Sap}_S(\beta)\!\downarrow(x)$. Thus,

$$\mathrm{Sap}_S(\alpha \cap \beta)\!\downarrow \supseteq \mathrm{Sap}_S(\alpha)\!\downarrow \cap \mathrm{Sap}_S(\beta)\!\downarrow. \qquad \text{(i)}$$

Now we show the reverse inclusion. We first note that $(\mathrm{Sap}_S(\alpha)\!\downarrow \cap \mathrm{Sap}_S(\beta)\!\downarrow)(x) = \mathrm{Sap}_S(\alpha)\!\downarrow(x) \cap \mathrm{Sap}_S(\beta)\!\downarrow(x) \leq \mathrm{Sap}_S(\alpha)\!\downarrow(x) \leq \alpha(y)$ for all $y \in N(x)$.
In a similar way, we have $\mathrm{Sap}_S(\alpha \cap \beta)\!\downarrow(x) \leq \mathrm{Sap}_S(\beta)\!\downarrow(x) \leq \beta(y)$ for all $y \in N(x)$. Thus, $(\mathrm{Sap}_S(\alpha)\!\downarrow \cap \mathrm{Sap}_S(\beta)\!\downarrow)(x) \leq \alpha(y) \cap \beta(y)$ for all $y \in N(x)$. It follows that $(\mathrm{IFSap}_S(\alpha)\!\downarrow \cap \mathrm{Sap}_S(\beta)\!\downarrow)(x) \leq \wedge\{\alpha(y) \cap \beta(y) : y \in N(x)\} = \mathrm{Sap}_S(\alpha \cap \beta)\!\downarrow(x)$. Thus,

$$\mathrm{Sap}_S(\alpha \cap \beta)\!\downarrow \subseteq \mathrm{Sap}_S(\alpha)\!\downarrow \cap \mathrm{Sap}_S(\beta)\!\downarrow. \qquad \text{(ii)}$$

From (i) and (ii), $\mathrm{Sap}_S(\alpha \cap \beta)\!\downarrow = \mathrm{Sap}_S(\alpha)\!\downarrow \cap \mathrm{Sap}_S(\beta)\!\downarrow$.

(g) Let $\alpha, \beta \in P(U)$, $x \in U$ and let $\alpha = \{(x, \mu_\alpha(x), v_\alpha(x) : x \in U)\}$, $\beta = \{(x, \mu_\beta(x), v_\beta(x) : x \in U)\}$. Put $N(x) = \{y : \exists e \in A/\{x, y\} \subseteq F(e)\}$. Now, $\alpha \cup \beta = \{(x, \mu_\alpha(x) \cup \mu_\beta(x), v_\alpha(x) \cap v_\beta(x) : x \in U)\}$. So $\mathrm{Sap}_S(\alpha)\!\downarrow(x) = \wedge \alpha(y) : y \in N(x)\} \leq \alpha(y) \leq \alpha(y) \cup \beta(y)$ for all $y \in N(x)$. Thus, $\mathrm{Sap}_S(\alpha \cup \beta)\!\downarrow(x) = \wedge\{\alpha(y) \cup \beta(y) : y \in N(x)\} \geq \mathrm{Sap}_S(\alpha)\!\downarrow(x)$. Similarly, we have $\mathrm{Sap}_S(\alpha \cup \beta)\!\downarrow(x) \geq \mathrm{Sap}_S(\beta)\!\downarrow(x)$.
Hence, it follows that $\mathrm{Sap}_S(\alpha \cup \beta)\!\downarrow(x) \geq \mathrm{Sap}_S(\alpha)\!\downarrow(x) \vee \mathrm{Sap}_S(\beta)\!\downarrow(x) = (\mathrm{Sap}_S(\alpha)\!\downarrow \cup \mathrm{Sap}_S(\beta)\!\downarrow)(x)$
Thus, we conclude that $\mathrm{Sap}_S(\alpha \cup \beta)\!\downarrow \supseteq \mathrm{Sap}_S(\alpha)\!\downarrow \cup \mathrm{Sap}_S(\beta)\!\downarrow$.

(h) The proof is similar to (f).

(i) The proof is similar to (g).

(j) Let $\alpha, \beta \in P(U)$, $x \in U$ and let $\alpha = \{(x, \mu_\alpha(x), v_\alpha(x) : x \in U)\}$, $\beta = \{(x, \mu_\beta(x), v_\beta(x) : x \in U)\}$. Put $N(x) = \{y : \exists e \in A/\{x, y\} \subseteq F(e)\}$.
If $\alpha \subseteq \beta \Rightarrow \mu_\alpha(x) \leq \mu_\beta(x)$ and $v_\alpha(x) \geq v_\beta(x)$.
Now, $\mathrm{Sap}_S(\alpha)\!\downarrow(x) = \wedge\{\alpha(y) : y \in N(x)\} \leq \alpha(y) \leq \beta(y)$ for all $y \in N(x)$. Then $\mathrm{Sap}_S(\alpha)\!\downarrow(x) \leq \wedge\{\beta(y) : y \in N(x)\} = \mathrm{Sap}_S(\beta)\!\downarrow(x)$.

(k) The proof is similar to (j). $\square$

Example 3.5 Suppose $U = \{h_1, h_2, h_3, h_4, h_5\}$ is the universe, consisting of five houses and set of parameters is given by $E = \{e_1, e_2, e_3, e_4, e_5\}$ where e_i ($i = 1, 2, 3, 4, 5$) stands for beautiful, modern, cheep, in green surrounding and in good repair, respectively. Let us consider a soft set (F, E) which describes the attractiveness of houses that Mr. Z is considering for purchase. In this case to define the soft set means to point out beautiful houses, modern houses and so on. Consider the mapping F given by 'houses (.)' where (.) is to be filled in by one parameter $e_i \in E$ for instance $F(e_1)$ means houses(beautiful) and its functional values is the set consisting of all the beautiful houses in U. Let $F(e_1) = h_5$, $F(e_2) = \{h_1, h_4\}$, $F(e_3) = \{h_1, h_2, h_3\}$ and $F(e_4) = \{h_3, h_5\}$. Let $\xi = (F, E)$ be the soft set over U. Let $S = \{U, \xi\}$ be a soft approximation space. Then, for intuitionistic fuzzy set $\alpha = \{h_1/(0.8, 0.2), h_2/(0.5, 0.3), h_3/(0.7, 0.3), h_4/(0.2, 0.7), h_5/(0.3, 0.6)\}$, we calculate $Sap_S(\alpha)\downarrow$ and $Sap_S(\alpha)\uparrow$ as follows:

$Sap_S(\alpha)\downarrow = \{h_1/(0.2, 0.7), h_2/(0.5, 0.3), h_3/(0.3, 0.6), h_4/(0.2, 0.7), h_5/(0.3, 0.6)\}$, and $Sap_S(\alpha)\uparrow = \{h_1/(0.8, 0.2), h_2/(0.8, 0.2), h_3/(0.8, 0.2), h_4/(0.8, 0.2), h_5/(0.7, 0.3)\}$. Similarly for the IF set, $\beta = \{h_1/(0.1, 0.8), h_2/(0.3, 0.6), h_3/(0.6, 0.4), h_4/(0.8, 0.2), h_5/(0.5, 0.5)\}$, we calculate $Sap_S(\beta)\downarrow$ and $Sap_S(\beta)\uparrow$ as follows $Sap_S(\beta)\downarrow = \{h_1/(0.1, 0.8), h_2/(0.1, 0.8), h_3/(0.1, 0.8), h_4/(0.1, 0.8), h_5/(0.5, 0.5)\}$, and $Sap_S(\beta)\uparrow = \{h_1/(0.8, 0.2), h_2/(0.6, 0.4), h_3/(0.6, 0.4), h_4/(0.8, 0.2), h_5/(0.6, 0.4)\}$. We first observe that

$$Sap_S(\alpha)\downarrow \subseteq Sap_S(\alpha)\uparrow \quad \text{and} \quad Sap_S(\beta)\downarrow \subseteq Sap_S(\beta)\uparrow.$$

Now, $(Sap_S(\alpha)\downarrow)^c = \{h_1/(0.7, 0.2), h_2/(0.3, 0.5), h_3/(0.6, 0.3), h_4/(0.7, 0.2), h_5/(0.6, 0.3)\}$, also $\alpha^c = \{h_1/(0.2, 0.8), h_2/(0.3, 0.5), h_3/(0.3, 0.7), h_4/(0.7, 0.2), h_5/(0.6, 0.3)\}$, so $Sap_S(\alpha^c)\uparrow = \{h_1/(0.7, 0.2), h_2/(0.3, 0.5), h_3/(0.6, 0.3), h_4/(0.7, 0.2), h_5/(0.6, 0.3)\}$. Thus,

$$(Sap_S(\alpha)\downarrow)^c = Sap_S(\alpha^c)\uparrow.$$

Now, $\alpha \cap \beta = \{h_1/(0.1, 0.8), h_2/(0.3, 0.6), h_3/(0.6, 0.4), h_4/(0.2, 0.7), h_5/(0.3, 0.6)\}$. We calculate $Sap_S(\alpha \cap \beta)\downarrow$ as follows:

$Sap_S(\alpha \cap \beta)\downarrow = \{h_1/(0.1, 0.8), h_2/(0.1, 0.8), h_3/(0.1, 0.8), h_4/(0.1, 0.8), h_5/(0.3, 0.6)\}$, and also we obtain $Sap_S(\alpha)\downarrow \cap Sap_S(\beta)\downarrow = \{h_1/(0.1, 0.8), h_2/(0.1, 0.8), h_3/(0.1, 0.8), h_4/(0.1, 0.8), h_5/(0.3, 0.6)\}$. Thus,

$$Sap_S(\alpha \cap \beta)\downarrow = Sap_S(\alpha)\downarrow \cap Sap_S(\beta)\downarrow.$$

Now, we calculate $Sap_S(\alpha \cap \beta)\uparrow$ as follows:

$Sap_S(\alpha \cap \beta)\uparrow = \{h_1/(0.6, 0.4), h_2/(0.6, 0.4), h_3/(0.6, 0.4), h_4/(0.2, 0.7), h_5/(0.6, 0.4)\}$. Also we obtain $Sap_S(\alpha)\uparrow \cap Sap_S(\beta)\uparrow = \{h_1/(0.8, 0.2), h_2/(0.6, 0.4), h_3/(0.6, 0.4), h_4/(0.8, 0.2), h_5/(0.6, 0.4)\}$.

Thus,

$$\mathrm{Sap}_S(\alpha \cap \beta)\uparrow \subseteq \mathrm{Sap}_S(\alpha)\uparrow \cap \mathrm{Sap}_S(\beta)\uparrow.$$

Now, $\alpha \cup \beta = \{h_1/(0.8, 0.2), h_2/(0.5, 0.3), h_3/(0.7, 0.3), h_4/(0.8, 0.2), h_5/(0.5, 0.5)\}$. We obtain, $\mathrm{IFSap}_S(\alpha \cup \beta)\downarrow = \{h_1/(0.5, 0.3), h_2/(0.5, 0.3), h_3/(0.5, 0.5), h_4/(0.8, 0.2), h_5/(0.5, 0.5)\}$. Also we calculate $\mathrm{Sap}_S(\alpha)\downarrow \cup \mathrm{Sap}_S(\beta)\downarrow = \{h_1/(0.2, 0.7), h_2/(0.5, 0.3), h_3/(0.3, 0.6), h_4/(0.2, 0.7), h_5/(0.5, 0.5)\}$.
Thus,

$$\mathrm{Sap}_S(\alpha \cup \beta)\downarrow \supseteq \mathrm{Sap}_S(\alpha)\downarrow \cup \mathrm{Sap}_S(\beta)\downarrow.$$

Now, $\mathrm{Sap}_S(\alpha \cup \beta)\uparrow = \{h_1/(0.8, 0.2), h_2/(0.8, 0.2), h_3/(0.8, 0.2), h_4/(0.8, 0.2), h_5/(0.7, 0.3)\}$. Also, $\mathrm{Sap}_S(\alpha)\uparrow \cup \mathrm{Sap}_S(\beta)\uparrow = \{h_1/(0.8, 0.2), h_2/(0.8, 0.2), h_3/(0.8, 0.2), h_4/(0.8, 0.2), h_5/(0.7, 0.3)\}$.
Thus,

$$\mathrm{Sap}_S(\alpha \cup \beta)\uparrow = \mathrm{Sap}_S(\alpha)\uparrow \cup \mathrm{Sap}_S(\beta)\uparrow.$$

Remark In this chapter, we first defined soft rough intuitionistic fuzzy sets (SRIF sets). Finally, we provided an example that demonstrated that this method can be successfully worked. It can be applied to problems of many fields that contain uncertainty. However, the approach should be more comprehensive in the future to solve the related problems. Now if $\alpha = \{(x, \mu_\alpha(x), \nu_\alpha(x) : x \in X)\}$ be an IF set, then taking $\nu_\alpha(x) = 1 - \mu_\alpha(x)$, it is clear that soft rough IF sets are soft rough fuzzy sets due to Feng et al. Thus, soft rough IF sets are the extension of soft rough fuzzy sets. Further, Feng et al. showed that soft rough fuzzy sets are the extension of rough fuzzy sets due to Dubois and Prade. Thus, our work is the extension of both the previous works of Dubois and Prade and Feng et al.

References

1. Aktas, H., Cagman, N.: Soft sets and soft groups. Inf. Sci. **1**(77), 2726–2735 (2007)
2. Atanassov, K.: Intuitionistic fuzzy sets. Fuzzy Sets Syst. **20**, 87–96 (1986)
3. Dubois, D., Prade, H.: Rough fuzzy sets and fuzzy rough sets. Int. J. Gen. Syst. **17**, 191–209 (1990)
4. Feng, F., Li, C., Davvaz, B., Ali, M.I.: Soft sets combined with fuzzy sets and rough sets: a tentative approach. Soft Comput. **14**, 899–911 (2010). doi:10.1007/s00500-009-0465-6
5. Maji, P.K., Biswas, R., Roy, A.R.: Fuzzy soft sets. J. Fuzzy Math. **9**(3), 589–602 (2001)
6. Molodtsov, D.A.: Soft set theory-first results. Comput. Math. Appl. **37**, 19–31 (1999)
7. Pawlak, Z.: Rough sets. Int. J. Inf. Comp. Sci. **11**, 341–356 (1982)
8. Zadeh, L.A.: Fuzzy sets. Inf. Control **8**, 338–353 (1965)

Chapter 4
Interval-Valued Intuitionistic Fuzzy Soft Rough Sets

Abstract In this chapter, the concept of interval-valued intuitionistic fuzzy soft rough sets is introduced. Also interval-valued intuitionistic fuzzy soft rough set-based multi-criteria group decision-making scheme is presented, which refines the primary evaluation of the whole expert group and enables us to select the optimal object in a most reliable manner. The proposed scheme is illustrated by an example regarding the candidate selection problem.

Keywords Soft set · Soft rough set · Interval-valued intuitionistic fuzzy set · Intuitionistic fuzzy soft rough set · Interval-valued intuitionistic fuzzy soft rough set · Decision-making

There are many complicated problems in economics, engineering, environmental science, and social science which cannot be solved by the well-known methods of classical mathematics as various types of uncertainties are presented in these problems. To overcome these uncertainties, some kinds of theories were given such as theory of fuzzy sets [14], rough sets [12], and soft sets [9] as mathematical tools for dealing with uncertainties. In 1999, Molodtsov [9] introduced soft set theory which is a new mathematical tool for dealing with uncertainties and is free from the difficulties affecting the existing methods. Research works on soft set theory are progressing rapidly. Combining soft sets with fuzzy sets and intuitionistic fuzzy sets, Maji et al. [7, 8] defined fuzzy soft sets and intuitionistic fuzzy soft sets which have rich potentials for solving decision-making problems. It has been found that soft set, fuzzy set, and rough set are closely related concepts. Based on the equivalence relation on the universe of discourse, Dubois and Prade [3] introduced the lower and upper approximation of fuzzy sets in a Pawlak s approximation space [12] and obtained a new notion called rough fuzzy sets. Feng et al. [4] introduced lower and upper soft rough approximations of fuzzy sets in a soft approximation space and obtained a new hybrid model called soft rough fuzzy sets which is the

© Springer India 2015
A. Mukherjee, *Generalized Rough Sets*, Studies in Fuzziness
and Soft Computing 324, DOI 10.1007/978-81-322-2458-7_4

extension of Dubois and Prade's rough fuzzy sets [3]. Considering lower and upper
intuitionistic fuzzy soft approximation space (IF soft approximation space),
Mukherjee [11] obtained a new hybrid model called intuitionistic fuzzy soft rough
set which can be seen as extension of both the previous works by Dubois and Prade
[3] and Feng et al. [4]. The notion of the interval-valued intuitionistic fuzzy set was
first introduced by Atanassov and Gargov [2]. It is characterised by an interval-
valued membership degree and an interval-valued non-membership degree. In
2010, Jiang et al. [6] introduced the concept of interval-valued intuitionistic fuzzy
soft sets. The aim of this chapter is to introduce a new concept—tinterval-valued
intuitionistic fuzzy soft rough sets. Also interval-valued intuitionistic fuzzy soft
rough set-based multi-criteria group decision-making scheme is to be presented.

Definition 4.1 Let $\Theta = (f, A)$ be a full soft set over U, i.e. $\cup_{a \in A} f(a) = U$ and
$S = (U, \Theta)$ be the soft approximation space. Then for $\tau \in \mathrm{IVIFS}^U$, the lower and
upper soft rough approximations of τ with respect to S are denoted by $\downarrow\mathrm{Sap}_S(\tau)$
and $\uparrow\mathrm{Sap}_S(\tau)$, respectively, which are interval-valued intuitionistic fuzzy sets in
U given by

$$\downarrow\mathrm{Sap}_S(\tau) = \{\langle x, [\wedge\{\inf \mu_\tau(y) : \exists a \in A(\{x, y\} \subseteq f(a))\},$$
$$\wedge\{\sup \mu_\tau(y) : \exists a \in (A\{x, y\} \subseteq f(a))\}],$$
$$[\vee\{\inf \gamma_\tau(y) : \exists a \in A(\{x, y\} \subseteq f(a))\},$$
$$\vee\{\sup \gamma_\tau(y) : \exists a \in A(\{x, y\} \subseteq f(a))\}]\rangle : x \in U\},$$

$$\uparrow\mathrm{Sap}_S(\tau) = \{\langle x, [\vee\{\inf \mu_\tau(y) : \exists a \in A(\{x, y\} \subseteq f(a))\},$$
$$\vee\{\sup \mu_\tau(y) : \exists a \in A(\{x, y\} \subseteq f(a))\}],$$
$$[\wedge\{\inf \gamma_\tau(y) : \exists a \in A(\{x, y\} \subseteq f(a))\},$$
$$\wedge\{\sup \gamma_\tau(y) : \exists a \in A(\{x, y\} \subseteq f(a))\}]\rangle : x \in U\}.$$

The operators and $\downarrow\mathrm{Sap}_S$ and $\uparrow\mathrm{Sap}_S$ are called the lower and upper soft rough
approximation operators on interval-valued intuitionistic fuzzy sets. If $\downarrow\mathrm{Sap}_S(\tau) =$
$\uparrow\mathrm{Sap}_S(\tau)$, then τ is said to be soft interval-valued intuitionistic fuzzy definable;
otherwise, τ is called an interval-valued intuitionistic fuzzy soft rough set.

Example 4.2 Let $U = \{x, y, z\}$ and $A = \{a, b, c\}$.
   Let $f : A \to P(U)$ be defined by $f(a) = \{x, y, z\}, f(b) = \{x, y\}, f(c) = \{x, z\}$.
   Let $\tau = \{\langle x, [0.3, 0.4], [0.1, 0.2]\rangle, \langle y, [0.6, 0.7], [0.1, 0.2]\rangle, \langle z, [0.5, 0.6], [0.2, 0.3]\rangle\}$.
Then, $\tau \in \mathrm{IVIFS}^U$.

So, we have,

$\downarrow \mathrm{Sap}_S(\tau) = \{\langle x, [0.5, 0.6], [0.1, 0.2]\rangle, \langle y, [0.3, 0.4], [0.1, 0.2]\rangle, \langle z, [0.3, 0.4], [0.2, 0.3]\rangle\}$
and
$\uparrow \mathrm{Sap}_S(\tau) = \{\langle x, [0.6, 0.7], [0.2, 0.3]\rangle, \langle y, [0.5, 0.6], [0.2, 0.3]\rangle, \langle z, [0.6, 0.7], [0.1, 0.2]\rangle\}.$

Since $\downarrow \mathrm{Sap}_S(\tau) \neq \uparrow \mathrm{Sap}_S(\tau)$, τ is an interval-valued intuitionistic fuzzy soft rough set.

Theorem 4.3 *Let $\Theta = (f, A)$ be a full soft set over U and $S = (U, \Theta)$ be the soft approximation space. Then for any $\tau \in \mathrm{IVIFS}^U$, we have*

(i) $\downarrow \mathrm{Sap}_S(\tau) = \{\langle x, [\wedge_{x \in f(a)} \wedge_{y \in f(a)} \inf \mu_\tau(y), \wedge_{x \in f(a)} \wedge_{y \in f(a)} \sup \mu_\tau(y)],$
     $[\vee_{x \in f(a)} \vee_{y \in f(a)} \inf \gamma_\tau(y), \vee_{x \in f(a)} \vee_{y \in f(a)} \sup \gamma_t(y)]\rangle : x \in U\},$

(ii) $\uparrow \mathrm{Sap}_S(\tau) = \{\langle x, [\vee_{x \in f(a)} \vee_{y \in f(a)} \inf \mu_\tau(y), \vee_{x \in f(a)} \vee_{y \in f(a)} \sup \mu_\tau(y)],$
     $[\wedge_{x \in f(a)} \wedge_{y \in f(a)} \inf \gamma_\tau(y), \wedge_{x \in f(a)} \wedge_{y \in f(a)} \sup \gamma_t(y)]\rangle : x \in U\}.$

Proof (i) Let $a \in A$ and $x \in f(a)$. Then for $y \in f(a)$, we have $\{x, y\} \subseteq f(a)$ and hence, $\inf \mu_\tau(y) \geq \wedge \{\inf \mu_\tau(z) : \exists a \in A(\{x, z\} \subseteq f(a))\}$. Consequently, $\wedge_{y \in f(a)} \inf \mu_\tau(y) \geq \wedge \{\inf \mu_\tau(z) : \exists a \in A(\{x, z\} \subseteq f(a))\}$ and so $\wedge_{x \in f(a)} \wedge_{y \in f(a)} \inf \mu_\tau(y) \geq \wedge \{\inf \mu_\tau(z) : \exists a \in A(\{x, z\} \subseteq f(a))\}$.

Similarly, it can be shown that

$$\wedge_{x \in f(a)} \wedge_{y \in f(a)} \sup \mu_\tau(y) \geq \wedge \{\sup \mu_\tau(z) : \exists a \in A(\{x, z\} \subseteq f(a))\}.$$

Thus, we get

$$[\wedge \{\inf \mu_\tau(z) : \exists a \in A(\{x, z\} \subseteq f(a))\}, \wedge \{\sup \mu_\tau(z) : \exists a \in A(\{x, z\} \subseteq f(a))\}]$$
$$\subseteq [\wedge_{x \in f(a)} \wedge_{y \in f(a)} \inf \mu_\tau(y), \wedge_{x \in f(a)} \wedge_{y \in f(a)} \sup \mu_\tau(y)]. \qquad (3.3.1)$$

In a similar manner, it can be shown that

$$[\vee \{\inf \gamma_\tau(z) : \exists a \in A(\{x, z\} \subseteq f(a))\}, \vee \{\sup \gamma_\tau(z) : \exists a \in A(\{x, z\} \subseteq f(a))\}]$$
$$\subseteq [\vee_{x \in f(a)} \vee_{y \in f(a)} \inf \gamma_\tau(y), \vee_{x \in f(a)} \vee_{y \in f(a)} \sup \gamma_\tau(y)]. \qquad (3.3.2)$$

From (3.3.1) and (3.3.2), we see that

$$\downarrow \mathrm{Sap}_S(\tau) \subseteq \{\langle x, [\wedge_{x \in f(a)} \wedge_{y \in f(a)} \inf \mu_\tau(y), \wedge_{x \in f(a)} \wedge_{y \in f(a)} \sup \mu_\tau(y)],$$
$$[\vee_{x \in f(a)} \vee_{y \in f(a)} \inf \gamma_\tau(y), \vee_{x \in f(a)} \vee_{y \in f(a)} \sup \gamma_\tau(y)] : x \in U\}. \qquad (3.3.3)$$

Now to prove that $\{\langle x, [\wedge_{x \in f(a)} \wedge_{y \in f(a)} \inf \mu_\tau(y), \wedge_{x \in f(a)} \wedge_{y \in f(a)} \sup \mu_\tau(y)], [\vee_{x \in f(a)} \vee_{y \in f(a)}], [\inf \gamma_\tau(y), \vee_{x \in f(a)} \vee_{y \inf(a)} \sup \gamma_\tau(y)]\rangle : x \in U\} \subseteq \downarrow \mathrm{Sap}_S(\tau)$, let us suppose that $a \in A$ such that $\{x, z\} \subseteq f(a)$. Then, $x \in f(a), z \in f(a)$, and hence,

$\inf \mu_\tau(z) \geq \wedge_{x \in f(a)} \wedge_{y \in f(a)} \inf \mu_\tau(y)$. Consequently, $\wedge\{\inf \mu_\tau(z) : \exists a \in A(\{x, z\} \subseteq f(a))\} \geq \wedge_{x \in f(a)} \wedge_{y \in f(a)} \inf \mu_\tau(y)$. Similarly, it can be shown that $\wedge\{\sup \mu_\tau(z): \exists a \in A(\{x, z\} \subseteq f(a))\} \geq \wedge_{x \in f(a)} \wedge_{y \in f(a)} \sup \mu_\tau(y)$. Thus, we get

$$\left[\wedge_{x \in f(a)} \wedge_{y \in f(a)} \inf \mu_\tau(y), \wedge_{x \in f(a)} \wedge_{y \in f(a)} \sup \mu_\tau(y)\right]$$
$$\subseteq [\wedge\{\inf \mu_\tau(z) : \exists a \in A(\{x, z\} \subseteq f(a))\},$$
$$\wedge\{\sup \mu_\tau(z) : \exists a \in A(\{x, z\} \subseteq f(a))\}]. \tag{3.3.4}$$

In a similar manner, it can be shown that

$$\left[\vee_{x \in f(a)} \vee_{y \in f(a)} \inf \gamma_\tau(y), \vee_{x \in f(a)} \vee_{y \in f(a)} \sup \gamma_\tau(y)\right]$$
$$\subseteq [\vee\{\inf \gamma_\tau(z) : \exists a \in A(\{x, z\} \subseteq f(a))\},$$
$$\vee\{\sup \gamma_\tau(z) : \exists a \in A(\{x, z\} \subseteq f(a))\}]. \tag{3.3.5}$$

From (3.3.4) and (3.3.5), we see that

$$\{\langle x, [\wedge_{x \in f(a)} \wedge_{y \in f(a)} \inf \mu_\tau(y), \wedge_{x \in f(a)} \wedge_{y \in f(a)} \sup \mu_\tau(y)],$$
$$[\vee_{x \in f(a)} \vee_{y \in f(a)} \inf \gamma_\tau(y), \vee_{x \in f(a)} \vee_{y \in f(a)} \sup \gamma_\tau(y)]\rangle : x \in U\} \subseteq \downarrow\text{Sap}_S(\tau) \tag{3.3.6}$$

From (3.3.3) and (3.3.6), we have $\downarrow\text{Sap}_S(\tau) = \{\langle x, [\wedge_{x \in f(a)} \wedge_{y \in f(a)} \inf \mu_\tau(y),$ $\wedge_{x \in f(a)} \wedge_{y \in f(a)} \sup \mu_\tau(y)], [\vee_{x \in f(a)} \vee_{y \in f(a)} \inf \gamma_\tau(y), \vee_{x \in f(a)} \vee_{y \in f(a)} \sup \gamma_\tau(y)]\rangle : x \in U\}$, (ii) is similar as in (i). □

Theorem 4.4 *Let $\Theta = (f, A)$ be a full soft set over U, $S = (U, \Theta)$ be the soft approximation space and $\tau, \delta \in \text{IVIFS}^U$. Then,*

(1) $\downarrow\text{Sap}_S(\phi) = \phi = \uparrow\text{Sap}_S(\phi)$
(2) $\downarrow\text{Sap}_S(U) = U = \uparrow\text{Sap}_S(U)$
(3) $\tau \subseteq \delta \Rightarrow \downarrow\text{Sap}_S(\tau) \subseteq \downarrow\text{Sap}_S(\delta)$
(4) $\tau \subseteq \delta \Rightarrow \uparrow\text{Sap}_S(\tau) \subseteq \uparrow\text{Sap}_S(\delta)$
(5) $\downarrow\text{Sap}_S(\tau \cap \delta) \subseteq \downarrow\text{Sap}_S(\tau) \cap \downarrow\text{Sap}_S(\delta)$
(6) $\uparrow\text{Sap}_S(\tau \cap \delta) \subseteq \uparrow\text{Sap}_S(\tau) \cap \uparrow\text{Sap}_S(\delta)$
(7) $\downarrow\text{Sap}_S(\tau) \cup \downarrow\text{Sap}_S(\delta) \subseteq \downarrow\text{Sap}_S(\tau \cup \delta)$
(8) $\uparrow\text{Sap}_S(\tau) \cup \uparrow\text{Sap}_S(\delta) \subseteq \uparrow\text{Sap}_S(\tau \cup \delta)$

Proof (1)–(4) are straight forward.
 (5) We have $\tau = \{\langle x, [\inf \mu_\tau(x), \sup \mu_\tau(x)], [\inf \gamma_\tau(x), \sup \gamma_\tau(x)]\rangle : x \in U\}$, $\delta = \{\langle x, [\inf \mu_\delta(x), \sup \mu_\delta(x)], [\inf \gamma_\delta(x), \sup \gamma_\delta(x)]\rangle : x \in U\}$, and $\tau \cap \delta = \{\langle x, [\inf \mu_{\tau \cap \delta}(x), \sup \mu_{\tau \cap \delta}(x)], [\inf \gamma_{\tau \cap \delta}(x), \sup \gamma_{\tau \cap \delta}(x)]\rangle : x \in U\}$. □

Now,

$$\downarrow \text{Sap}_S(\tau \cap \delta) = \{\langle x, [\wedge\{\inf \mu_{\tau \cup \delta}(y) : \exists a \in A(\{x,y\} \subseteq f(a))\},$$
$$\wedge\{\sup \mu_{\tau \cap \delta}(y) : \exists a \in A(\{x,y\} \subseteq f(a))\}],$$
$$[\vee\{\inf \gamma_{\tau \cup \delta}(y) : \exists a \in A(\{x,y\} \subseteq f(a))\},$$
$$\vee\{\sup \gamma_{\tau \cup \delta}(y) : \exists a \in A(\{x,y\} \subseteq f(a))\}]\rangle : x \in U\}$$
$$= \{\langle x, [\wedge\{\min(\inf \mu_{\tau}(y), \inf \mu_{\delta}(y)) : \exists a \in A(\{x,y\} \subseteq f(a))\},$$
$$\wedge\{\min(\sup \mu_{\tau}(y), \sup \mu_{\delta}(y)) : \exists a \in A(\{x,y\} \subseteq f(a))\}],$$
$$[\vee\{\max(\inf \gamma_{\tau}(y), \inf \gamma_{\delta}(y)) : \exists a \in A(\{x,y\} \subseteq f(a))\},$$
$$\vee\{\max(\sup \gamma_{\tau}(y), \sup \gamma_{\delta}(y) : \exists a \in A(\{x,y\} \subseteq f(a))\}]\rangle : x \in U\}.$$

$$(3.3.1)$$

Since $\min(\inf \mu_{\tau}(y), \inf \mu_{\delta}(y)) \leq \inf \mu_{\tau}(y)$ and $\min(\inf \mu_{\tau}(y), \inf \mu_{\delta}(y)) \leq \inf \mu_{\delta}(y)$ we have $\wedge\{\min(\inf \mu_{\tau}(y), \inf \mu_{\delta}(y)) : \exists a \in A(\{x,y\} \subseteq f(a))\} \leq \wedge\{\inf \mu_{\tau}(y) : \exists a \in A(\{x,y\} \subseteq f(a))\}$ and $\wedge\{\min(\inf \mu_{\tau}(y), \inf \mu_{\delta}(y)) : \exists a \in A(\{x,y\} \subseteq f(a))\}$ $\leq \wedge\{\inf \mu_{\delta}(y) : \exists a \in A(\{x,y\} \subseteq f(a))\}$. Consequently,

$$\wedge\{\min(\inf \mu_{\tau}(y), \inf \mu_{\delta}(y)) : \exists a \in A(\{x,y\} \subseteq f(a))\} \leq \min(\wedge\{\inf \mu_{\tau}(y) :$$
$$\exists a \in A(\{x,y\} \subseteq f(a))\}, \wedge\{\inf \mu_{\delta}(y) : \exists a \in A(\{x,y\} \subseteq f(a))\}). \qquad (3.3.2)$$

Similarly, we can get

$$\wedge\{\min(\sup \mu_{\tau}(y), \sup \mu_{\delta}(y)) : \exists a \in A(\{x,y\} \subseteq f(a))\} \leq \min(\wedge\{\sup \mu_{\tau}(y) :$$
$$\exists a \in A(\{x,y\} \subseteq f(a))\}, \wedge\{\sup \mu_{\delta}(y) : \exists a \in A(\{x,y\} \subseteq f(a))\}).$$

$$(3.3.3)$$

Again since $\max(\inf \gamma_{\tau}(y), \inf \gamma_{\delta}(y)) \geq \inf \gamma_{\tau}(y)$ and $\max(\inf \gamma_{\tau}(y), \inf \gamma_{\delta}(y)) \geq \inf \gamma_{\delta}(y)$, we have,

$$\vee\{\max(\inf \gamma_{\tau}(y), \inf \gamma_{\delta}(y)) : \exists a \in A(\{x,y\} \subseteq f(a))\}$$
$$\geq \vee\{\inf \gamma_{\tau}(y) : \exists a \in A(\{x,y\} \subseteq f(a))\} \text{ and}$$
$$\vee\{\max(\inf \gamma_{\tau}(y), \inf \gamma_{\delta}(y)) : \exists a \in A(\{x,y\} \subseteq f(a))\}$$
$$\geq \vee\{\inf \gamma_{\delta}(y) : \exists \in A(\{x,y\} \subseteq f(a))\}.$$

Consequently,

$$\vee\{\max(\inf \gamma_{\tau}(y), \inf \gamma_{\delta}(y)) : \exists a \in A(\{x,y\} \subseteq f(a))\} \geq \max(\vee\{\inf \gamma_{\tau}(y) :$$
$$\exists a \in A(\{x,y\} \subseteq f(a))\}, \vee\{\inf \gamma_{\delta}(y) : \exists a \in A(\{x,y\} \subseteq f(a))\}). \qquad (3.3.4)$$

Similarly, we can get

$$\vee\{\max(\sup \gamma_\tau(y), \sup \gamma_\delta(y)) : \exists a \in A(\{x,y\} \subseteq f(a))\} \geq \max(\vee\{\sup \gamma_\tau(y) : \\ \exists a \in A(\{x,y\} \subseteq f(a))\}, \vee\{\sup \gamma_\delta(y) : \exists a \in A(\{x,y\} \subseteq f(a))\}). \tag{3.3.5}$$

Using (3.3.2)–(3.3.5), we get from (3.3.1) $\downarrow Sap_S(\tau \cap \delta) \subseteq \downarrow Sap_S(\tau) \cap Sap_S(\delta)$.

(6) This proof is similar to (5).

(7) We have $\tau = \{\langle x, [\inf \mu_\tau(x), \sup \mu_\tau(x)], [\inf \gamma_\tau(x), \sup \gamma_\tau(x)]\rangle : x \in U\}$, $\delta = \{\langle x, [\inf \mu_\delta(x), \sup \mu_\delta(x)], [\inf \gamma_\delta(x), \sup \gamma_\delta(x)]\rangle : x \in U\}$ and $\tau \cup \delta = \{\langle x, [\inf \mu_{\tau\cup\delta}(x), \sup \mu_{\tau\cup\delta}(x)], [\inf \gamma_{\tau\cup\delta}(x), \sup \gamma_{\tau\cup\delta}(x)]\rangle : x \in U\}$.

Now,

$$\downarrow Sap_S(\tau \cup \delta) = \{\langle x, [\wedge\{\inf \mu_{\tau\cup\delta}(y) : \exists a \in A(\{x,y\} \subseteq f(a))\}, \\ \wedge\{\sup \mu_{\tau\cup\delta}(y) : \exists a \in A\{x,y\} \subseteq f(a))\}], \\ [\vee\{\inf \gamma_{\tau\cup\delta}(y) : \exists a \in A(\{x,y\} \subseteq f(a))\}, \\ \vee\{\sup \gamma_{\tau\cup\delta}(y) : \exists a \in A(\{x,y\} \subseteq f(a))\}]\rangle : x \in U\} \\ \{\langle x, [\wedge\{\max(\inf \mu_\tau(y), \inf \mu_\delta(y)) : \exists a \in A(\{x,y\} \subseteq f(a))\}, \\ \wedge\{\max(\sup \mu_\tau(y), \sup \mu_\delta(y)) : \exists a \in A(\{x,y\} \subseteq f(a))\}, \\ [\vee\{\min(\inf \gamma_\tau(y), \inf \gamma_\delta(y)) : \exists a \in A(\{x,y\} \subseteq f(a))\}, \\ \vee\{\min(\sup \gamma_\tau(y), \sup \gamma_\delta(y) : \exists a \in A(\{x,y\} \subseteq f(a))\}: x \in U\}. \tag{3.3.6}$$

Since $\max(\inf \mu_\tau(y), \inf \mu_\delta(y)) \leq \inf \mu_\tau(y)$ and $\max(\inf \mu_\tau(y), \inf \mu_\delta(y)) \leq \inf \mu_\delta(y)$, we have $\wedge\{\max(\inf \mu_\tau(y), \inf \mu_\delta(y)) : \exists a \in A(\{x,y\} \subseteq f(a))\} \geq \wedge\{\inf \mu_\tau(y) : \exists a \in A(\{x,y\} \subseteq f(a))\}$ and $\wedge\{\max(\inf \mu_\tau(y), \inf \mu_\delta(y)) : \exists a \in A(\{x,y\} \subseteq f(a))\} \geq \wedge\{\inf \mu_\delta(y) : \exists a \in A(\{x,y\} \subseteq f(a))\}$.

Consequently,

$$\wedge\{\max(\inf \mu_\tau(y), \inf \mu_\delta(y)) : \exists a \in A(\{x,y\} \subseteq f(a))\} \geq \max(\wedge\{\inf \mu_\tau(y) : \\ \exists a \in A(\{x,y\} \subseteq f(a))\} \wedge \{\inf \mu_\delta(y) : \exists a \in A(\{x,y\} \subseteq f(a))\}). \tag{3.3.7}$$

Similarly, we can get

$$\wedge\{\max(\sup \mu_\tau(y), \sup \mu_\delta(y)) : \exists a \in A(\{x, y\} \subseteq f(a))\} \geq \max\wedge(\sup \mu_\tau(y) : \\ \exists a \in A(\{x,y\} \subseteq f(a))\}, \wedge\{\sup \mu_\delta(y) : \exists a \in A(\{x,y\} \subseteq f(a))\}). \tag{3.3.8}$$

Again since $\min(\inf \gamma_\tau(y), \inf \gamma_\delta(y)) \leq \inf \gamma_\tau(y)$ and $\min(\inf \gamma_\tau(y), \inf \gamma_\delta(y)) \leq \inf \gamma_\delta(y)$, we have,

$$\vee\{\min(\inf \gamma_\tau(y), \inf \gamma_\delta(y)) : \exists a \in A(\{x,y\} \subseteq f(a))\} \\ \leq \vee\{\inf \gamma_\tau(y) : \exists a \in A(\{x,y\} \subseteq f(a))\}$$

and

$$\vee\{\min(\inf \gamma_\tau(y),\ \inf \gamma_\delta(y)) : \exists a \in A(\{x,y\} \subseteq f(a))\}$$
$$\leq \vee\{\inf \gamma_\tau(y) : \exists a \in A(\{x,y\} \subseteq f(a))\}.$$

Consequently,

$$\vee\{\min(\inf \gamma_\delta(y), \inf \gamma_\delta(y)) : \exists a \in A(\{x,y\} \subseteq f(a))\} \leq \min(\vee\{\inf \gamma_\tau(y) :$$
$$\exists a \in A(\{x,y\} \subseteq f(a))\}, \vee\{\inf \gamma_\delta(y) : \exists a \in A(\{x,y\} \subseteq f(a))\}). \qquad (3.3.9)$$

Similarly we can get,

$$\vee\{\min(\sup \gamma_\tau(y),\ \sup \gamma_\delta(y)) : \exists a \in A(\{x,y\} \subseteq f(a))\} \leq \min(\vee\{\sup \gamma_\tau(y) :$$
$$\exists a \in A(\{x,y\} \subseteq f(a))\}, \vee\{\sup \gamma_\delta(y) : \exists a \in A(\{x,y\} \subseteq f(a))\}). \qquad (3.3.10)$$

Using (3.3.7)–(3.3.10), we get from (3.3.6), $\downarrow \mathrm{Sap}_S(\tau) \cup \downarrow \mathrm{Sap}_S(\delta) \subseteq \downarrow \mathrm{Sap}_S(\tau \cup \delta)$.
(8) This proof is similar to (7).

4.1 A Multi-Criteria Group Decision-Making Problem

Soft sets, fuzzy soft sets, and intuitionistic fuzzy soft sets have been applied by many authors in solving decision-making problems. In this section, we illustrate the use of soft sets and fuzzy soft sets, intuitionistic fuzzy soft sets, interval-valued intuitionistic fuzzy soft sets, rough sets, interval-valued intuitionistic fuzzy soft rough sets, and related notions in object evaluation and group decision-making.

Let $U = \{o_1, o_2, o_3, \ldots, o_r\}$ be a set of objects and E be a set of parameters and $A = \{e_1, e_2, e_3, \ldots, e_m\} \subseteq E$ and $S = (F, A)$ be a full soft set over U. Let us assume that we have an expert group $G = \{T_1, T_2, , T_n\}$ consisting of n specialists to evaluate the objects in U. Each specialist will examine all the objects in U and will point out his/her evaluation result. Let X_i denote the primary evaluation result of the specialist T_i. It is easy to see that the primary evaluation result of the whole expert group G can be represented as an interval-valued intuitionistic fuzzy evaluation soft set $S^* = (F^*, G)$ over U where $F^* : G \to \mathrm{IVIFS}^U$ is given by $F^*(T_i) = X_i$, for $i = 1, 2, \ldots, n$.

Now we consider the soft rough interval-valued intuitionistic fuzzy approximations of the specialist T_i's primary evaluation result X_i with respect to the soft approximation space $P = (U, S)$. Then, we obtain two other interval-valued intuitionistic fuzzy soft sets $\downarrow S^* = (\downarrow F^*, G)$ and $\downarrow S^* = (\downarrow F^*, G)$ over U where $\downarrow F^* : G \to \mathrm{IVIFS}^U$ is given by $\downarrow F^*(T_i) = \downarrow \mathrm{apr}_P(X_i)$ and $\uparrow F^* : G \to \mathrm{IVIFS}^U$ is given by $\uparrow F^*(T_i) = \uparrow \mathrm{apr}_P(X_i)$, for $i = 1, 2, \ldots, n$.

Here, $\downarrow S^*$ can be considered as the evaluation result for the whole expert group G with 'low confidence', $\uparrow S^*$ can be considered as the evaluation result for the whole expert group G with 'high confidence', and S^* can be considered as the evaluation result for the whole expert group G with 'middle confidence'.

Let us define two interval-valued intuitionistic fuzzy sets IVIFSet$_{\downarrow S^*}$ and IVIFSet$_{\downarrow S^*}$ by

$$
\text{IVIFSet}_{\downarrow S^*} = \left\{ \left\langle o_k, \left[\frac{1}{n} \sum_{j=1}^{n} \inf \mu \downarrow_{F^*(T_j)} (o_k), \frac{1}{n} \sum_{j=1}^{n} \sup \mu \downarrow_{F^*(T_j)} (o_k) \right], \right. \right.
$$

$$
\left. \left. \left[\frac{1}{n} \sum_{j=1}^{n} \inf \gamma \downarrow_{F^*(T_j)} (o_k), \frac{1}{n} \sum_{j=1}^{n} \sup \gamma \downarrow_{F^*(T_j)} (o_k) \right] \right\rangle \right.
$$

$$
\left. : k = 1, 2, \ldots, l \right\} \text{ and}
$$

$$
\text{IVIFSet}_{\uparrow S^*} = \left\{ \left\langle o_k, \left[\frac{1}{n} \sum_{j=1}^{n} \inf \mu \uparrow_{F^*(T_j)} (o_k), \frac{1}{n} \sum_{j=1}^{n} \sup \mu \uparrow_{F^*(T_j)} (o_k) \right], \right. \right.
$$

$$
\left. \left. \left[\frac{1}{n} \sum_{j=1}^{n} \inf \gamma \uparrow_{F^*(T_j)} (o_k), \frac{1}{n} \sum_{j=1}^{n} \sup \gamma \uparrow_{F^*(T_j)} (o_k) \right] \right\rangle \right.
$$

$$
\left. : k = 1, 2, \ldots, l \right\}.
$$

Now we define another interval-valued intuitionistic fuzzy set IVIFSet$_{S^*}$ by

$$
\text{IVIFSet}_{S^*} = \left\{ \left\langle o_k, \left[\frac{1}{n} \sum_{j=1}^{n} \inf \mu_{F^*(T_j)}(o_k), \frac{1}{n} \sum_{j=1}^{n} \sup \mu_{F_*(T_j)}(o_k) \right], \right. \right.
$$

$$
\left. \left. \left[\frac{1}{n} \sum_{j=1}^{n} \inf \gamma_{F_*(T_j)}(o_k), \frac{1}{n} \sum_{j=1}^{n} \sup \gamma_{F^*(T_j)}(o_k) \right] \right\rangle : k = 1, 2, \ldots, l \right\}
$$

Then clearly, IVIFSet$_{\downarrow S^*} \subseteq$ IVIFSet$_{S^*} \subseteq$ IVIFSet$_{\uparrow S^*}$.

Let $C = \{L(\text{low confidence}), M(\text{middle confidence}), H(\text{high confidence})\}$ be a set of parameters. Let us consider the interval-valued intuitionistic fuzzy soft set $S^{**} = (f, C)$ over U where $f : C \rightarrow \text{IVIFS}^U$ is given by

$$
f(L) = \text{IVIFSet}_{\downarrow S^*}, f(M) = \text{IVIFSet}_{S^*}, f(H) = \text{IVIFSet}_{\uparrow} S^*.
$$

Now given a weighting vector $W = (w_L, w_M, w_H)$ such that $w_L, w_M, w_H \in \text{Int}([0, 1])$, we define $\alpha : U \rightarrow P(U)$ by

$$
\alpha(o_k) = \sup w_L * \sup \mu_{f(L)}(o_k) + \sup w_M * \sup \mu_{f(M)}(o_k)
$$

$$
+ \sup w_H * \sup \mu_{f(H)}(o_k), o_k \in U
$$

$$
(* \text{ represents ordinary multiplication}).
$$

Here $\alpha(o_k)$ is called the weighted evaluation value of the alternative $o_k \in U$. Finally, we can select the object o_p such that $\alpha(o_p) = \max\{\alpha(o_k) : k = 1, 2, \ldots, r\}$ as the most preferred alternative.

Algorithm

(1) Input the original description soft set (F, A).
(2) Construct the interval-valued intuitionistic fuzzy evaluation soft set $S^* = (F^*, G)$
(3) Compute the soft rough interval-valued intuitionistic fuzzy approximations and then construct the interval-valued intuitionistic fuzzy soft sets $\downarrow S^*$ and $\uparrow S^*$.
(4) Construct the interval-valued intuitionistic fuzzy sets IVIFSet$_{\downarrow S^*}$, IVIFSet$_{S^*}$, IVIFSet$_{\downarrow S^*}$.
(5) Construct the interval-valued intuitionistic fuzzy soft set S^{**}.
(6) Input the weighting vector W and compute the weighted evaluation values $\alpha(o_k)$ of each alternative $o_k \in U$.
(7) Select the object o_p such that $\alpha(o_p) = \max\{\alpha(o_k) : k = 1, 2, \ldots, r\}$ as the most preferred alternative.

An Illustrative Example

Let us consider a staff selection problem to fill a position in a private company. Let $U = \{c_1, c_2, c_3, c_4, c_5\}$ is the universe set consisting of five candidates. Let us consider the soft set $S = (F, A)$, which describes the quality of the candidates where $A = \{e_1(\text{experience}), e_2(\text{computer knowledge}), e_3(\text{young and efficient}), e_4(\text{good communication skill})\}$.

Let the tabular representation of the soft set (F, A) be:

	c_1	c_2	c_3	c_4	c_5
e_1	1	0	1	1	0
e_2	1	1	0	1	0
e_3	0	1	1	1	1
e_4	1	1	0	0	1

Let $G = \{T_1, T_2, T_3, T_4\}$ be the set of interviewers to judge the quality of the candidate in U. Now if X_i denote the primary evaluation result of the interviewer T_i (for $i = 1, 2, 3, 4$), then the primary evaluation result of the whole expert group G can be represented as an interval-valued intuitionistic fuzzy evaluation soft set $S^* = (F^*, G)$ over U where $F^*: G \rightarrow \text{IVIFS}^U$ is given by $F^*(T_i) = X_i$ for $i = 1, 2, 3, 4$.
Let the tabular representation of S^* be given as

	c_1	c_2	c_3	c_4	c_5
T_1	([0.2, 0.3], [0.4, 0.5])	([0.6, 0.7], [0.1, 0.2])	([0.3, 0.4], [0.3, 0.5])	([0.2, 0.4], [0.4, 0.6])	([0.3, 0.6], [0.2, 0.3])
T_2	([0.1, 0.3], [0.6, 0.7])	([0.3, 0.4], [0.4, 0.5])	([0.5, 0.7], [0.1, 0.2])	([0.7, 0.8], [0.1, 0.2])	([0.1, 0.3], [0.1, 0.5])
T_3	([0.4, 0.6], [0.2, 0.3])	([0.1, 0.4], [0.2, 0.4])	([0.2, 0.5], [0.2, 0.4])	([0.3, 0.5], [0.2, 0.4])	([0.4, 0.5], [0.2, 0.5])
T_4	([0.3, 0.5], [0.3, 0.4])	([0.5, 0.6], [0.2, 0.3])	([0.4, 0.5], [0.2, 0.5])	([0.4, 0.7], [0.1, 0.2])	([0.6, 0.8], [0.1, 0.2])

Let us choose $P = (U, S)$ as the soft interval-valued intuitionistic fuzzy approximation space. Let us consider the interval-valued intuitionistic fuzzy evaluation soft sets $\downarrow S^* = (\downarrow F^*, G)$ and $\uparrow S^* = (\uparrow F^*, G)$ over U.

Then, the tabular representation of these sets are as follows:

$(\downarrow F^*, G)$:

	c_1	c_2	c_3	c_4	c_5
T_1	([0.2, 0.4], [0.4, 0.6])	([0.2, 0.3], [0.4, 0.6])	([0.2, 0.3], [0.4, 0.6])	([0.2, 0.3], [0.4, 0.6])	([0.2, 0.3], [0.4, 0.6])
T_2	([0.1, 0.2], [0.6,0.8])	([0.1, 0.3], [0.6, 0 .7])	([0.1, 0.3], [0.6, 0.7])	([0.1, 0.3], [0.6, 0.7])	([0.1, 0.3], [0.6, 0.7])
T_3	([0.1, 0.4], [0.2, 0.5])	([0.1, 0.2], [0.2, 0.5])	([0.1, 0.4], [0.2, 0.5])	([0.1, 0.4], [0.2, 0.5])	([0.1, 0.4], [0.2, 0.6])
T_4	([0.2, 0.4], [0.4, 0.5])	([0.3, 0.5], [0.3, 0.5])	([0.3, 0.5], [0.3, 0.5])	([0.3, 0.5], [0.3, 0.5])	([0.3, 0.5], [0.3, 0.5])

$(\uparrow F^*, G)$:

	c_1	c_2	c_3	c_4	c_5
T_1	([0.6, 0.7], [0.1, 0.2])	([0.6, 0.8], [0.1, 0.2])	([0.6, 0.7], [0.1, 0.2])	([0.6, 0.7], [0.1, 0.2])	([0.6, 0.7], [0.1, 0.2])
T_2	([0.7, 0.8], [0.1, 0.2])	([0.7, 0.8], [0.1, 0.2])	([0.7, 0.8], [0.1, 0.2])	([0.5, 0.7], [0.1, 0.2])	([0.7, 0.8], [0.1, 0.2])
T_3	([0.4, 0.7], [0.2, 0.3])	([0.4, 0.6], [0.2, 0.3])	([0.4, 0.6], [0.2, 0.3])	([0.4, 0.6], [0.2, 0.3])	([0.4, 0.6], [0.2, 0.3])
T_4	([0.6, 0.8], [0.1, 0.2])	([0.6, 0.8], [0.1, 0.2])	([0.6, 0.8], [0.1, 0.2])	([0.6, 0.8], [0.1, 0.2])	([0.5, 0.7], [0.1, 0.2])

Here, $\downarrow S^* \subseteq S^* \subseteq \uparrow S^*$

Then, we have

$$\text{IVIFSet}_{\downarrow S^*} = \{\langle c_1, [0.15, 0.35], [0.4, 0.625]\rangle, \langle c_2, [0.175, 0.325], [0.375, 0.575]\rangle,$$
$$\langle c_3, [0.175, 0.375], [0.375, 0.575]\rangle, \langle c_4, [0.175, 0.375], [0.375, 0.575]\rangle,$$
$$\langle c_5, [0.175, 0.375], [0.375, 0.6]\rangle\},$$

$\text{IVIFSet}_{\uparrow S^*} = \{\langle c_1, [0.575, 0.75], [0.125, 0.225]\rangle, \langle c_2, [0.575, 0.75], [0.125, 0.225]\rangle,$
$\qquad \langle c_3, [0.575, 0.725], [0.125, 0.225]\rangle, \langle c_4, [0.525, 0.700], [0.125, 0.225],$
$\qquad \langle c_5, [0.55, 0.700], [0.125, 0.225]\rangle\},$

$\text{IVIFSet}_{S^*} = \{\langle c_1, [0.25, 0.425], [0.375, 0.475]\rangle, \langle c_2, [0.375, 0.525], [0.225, 0.35]\rangle,$
$\qquad \langle c_3, [0.350, 0.525], [0.2, 0.4]\rangle, \langle c_4, [0.4, 0.6], [0.20, 0.35]\rangle,$
$\qquad \langle c_5, [0.35, 0.55], [0.15, 0.375]\rangle\}.$

Here, $\text{IVIFSet}_{\downarrow S^*} \subseteq \text{IVIFSet}_{S^*} \subseteq \text{IVIFSet}_{\uparrow S^*}$.

Let $C = \{L(\text{low confidence}), M(\text{middle confidence}), H(\text{high confidence})\}$ be a set of parameters. Let us consider the interval-valued intuitionistic fuzzy soft set $S^{**} = (f, C)$ over U where $f: C \rightarrow \text{IVIFS}^U$ is given by

$$f(L) = \text{IVIFSet}_{\downarrow S^*}, f(M) = \text{IVIFSet}_{S^*}, f(H) = \text{IVIFSet}_{\uparrow S^*}$$

Now assuming the weighting vector $W = (w_L, w_M, w_H)$ such that $w_L = [0.5, 0.7]$, $w_M = [0.4, 0.6]$, $w_H = [0.4, 0.8]$, we have

$$\alpha(c_1) = 0.7 \times 0.35 + 0.6 \times 0.425 + 0.8 \times 0.75 = 1.100,$$
$$\alpha(c_2) = 0.7 \times 0.325 + 0.6 \times 0.525 + 0.8 \times 0.75 = 1.425,$$
$$\alpha(c_3) = 0.7 \times 0.375 + 0.6 \times 0.525 + 0.8 \times 0.725 = 1.1575,$$
$$\alpha(c_4) = 0.7 \times 0.375 + 0.6 \times 0.6 + 0.8 \times 0.7 = 1.1825,$$
$$\alpha(c_5) = 0.7 \times 0.375 + 0.6 \times 0.55 + 0.8 \times 0.7 = 1.1525.$$

Since $\max\{\alpha(c_1), \alpha(c_2), \alpha(c_3), \alpha(c_4), \alpha(c_5)\} = 1.1825$, so the candidate c_4 will be selected as the most preferred alternative.

4.2 Conclusion

In this chapter, we first defined interval-valued intuitionistic fuzzy soft rough sets (IVIFSsets). Finally, we provided an example that demonstrated that this method can be successfully worked. It can be applied to the problems of many fields that contain uncertainty. However, the approach should be more comprehensive in the future to solve the related problems. It is clear that IVIF soft rough sets are IF soft rough sets due to Mukherjee. Also IF soft rough sets are soft rough fuzzy sets due to Feng et al. Further, Feng et al. showed that soft rough fuzzy sets are the extension of rough fuzzy sets due to Dubois and Prade. Thus, our work is the extension of the previous works of Mukherjee, Dubois, Prade, and Feng et al.

References

1. Atanassov, K.: Intuitionistic fuzzy sets. Fuzzy Sets Syst. **20**, 87–96 (1986)
2. Atanassov, K., Gargov, G.: Interval-valued intuitionistic fuzzy sets. Fuzzy Sets Syst. **31**, 343–349 (1989)
3. Dubois, D., Prade, H.: Rough fuzzy sets and fuzzy rough sets. Int. J. Gen Syst. **17**, 191–209 (1990)
4. Feng, F., Li, C.X., Davvaz, B., Ali, M.I.: Soft sets combined with fuzzy sets and rough sets: a tentative approach. Soft. Comput. **14**, 899–911 (2010)
5. Feng, F.: Soft rough sets applied to multi criteria group decision making. Ann. Fuzzy Math. Inf. **2**, 69–80 (2011)
6. Jiang, Y., Tang, Y., Chen, Q., Liu, H., Tung, J.: Interval-valued intuitionistic fuzzy soft sets and their properties. Comput. Math. Appl. **60**, 906–918 (2010)
7. Maji, P.K., Biswas, R., Roy, A.R.: Fuzzy soft sets. J. Fuzzy Math. **9**, 589–602 (2001)
8. Maji, P.K., Biswas, R., Roy, A.R.: Intuitionistic fuzzy soft sets. J. Fuzzy Math. **9**(3), 677–692 (2001)
9. Molodtsov, D.: Soft set theory-first results. Comput. Math. Appl. **37**(4–5), 19–31 (1999)
10. Meng, D., Zhang, X., Qin, K.: Soft rough fuzzy sets and soft fuzzy rough sets. Comput. Math. Appl. **62**, 4635–4645 (2011)
11. Mukherjee, A.: Intuitionistic fuzzy soft rough sets. J. Fuzzy Math. V20(1) 2012 (to appear)
12. Pawlak, Z.: Rough sets. Int. J. Comput. Inf. Sci. **11**, 341–356 (1982)
13. Yang, X.B., Lin, T.Y., Yang, J.Y., Li, Y., Yu, D.J.: Combination of interval-valued fuzzy set and soft set. Comput. Math. Appl. **58**, 521–527 (2009)
14. Zadeh, L.A.: Fuzzy sets. Inf. Control **8**, 338–353 (1965)

Chapter 5
Interval-Valued Intuitionistic Fuzzy Soft Topological Spaces

Abstract In this chapter, the concept of interval-valued intuitionistic fuzzy soft topological space (IVIFS topological space) together with intuitionistic fuzzy soft open sets (IVIFS open sets) and intuitionistic fuzzy soft closed sets (IVIFS closed sets) are introduced. We define neighbourhood of an IVIFS set, interior IVIFS set, interior of an IVIFS set, exterior IVIFS set, exterior of an IVIFS set, closure of a IVIFS set, IVIFS basis, and IVIFS subspace. Some examples and theorems regarding these concepts are presented.

Keywords IVIFS topology · Neighbourhood of an IVIFS set · Interior IVIFS set · Interior of an IVIFS set · Exterior IVIFS set · Exterior of an IVIFS set · Closure of a IVIFS set · IVIFS basis · IVIFS subspace

The vagueness or the representation of imperfect knowledge has been a problem for a long time for the mathematicians. There are many mathematical tools for dealing with uncertainties; some of them are fuzzy set theory [13] and soft set theory [8]. Maji et al. [5] defined several operations on soft set theory. Based on the analysis of several operations on soft sets introduced in [5], Ali et al. [1] presented some new algebraic operations for soft sets and proved that certain De Morgan's law holds in soft set theory with respect to these new definitions. Combining soft sets with fuzzy sets [13] and intuitionistic fuzzy sets [2], Maji et al. [6, 7] defined fuzzy soft sets and intuitionistic fuzzy soft sets which are rich potentials for solving decision-making problems. As a generalisation of fuzzy soft set theory, intuitionistic fuzzy soft set theory makes description of the objective more realistic, more practical, and accurate in some cases, making it more promising. In 2011, Shabir and Naz [10] defined soft topology by using soft sets and presented basic properties in their paper. Tanay and Kandemir [12] defined fuzzy soft topology on a fuzzy soft set over an initial universe. Later on, Simsekler and Yuksel [11] introduced fuzzy soft topology over a fuzzy soft set with a fixed parameter set. Zhaowen et al. [14] defined the topological structure of intuitionistic fuzzy soft sets taking the whole parameter set E. In this chapter, we introduce the concepts of intuitionistic fuzzy soft topological space (IVIFS topological space) together with intuitionistic fuzzy soft open sets (IVIFS open sets) and intuitionistic fuzzy soft closed sets (IVIFS closed sets). Then, we define neighbourhood of an IVIFS set, interior IVIFS set, interior of an IVIFS set, exterior IVIFS set, exterior of an IVIFS set, closure of an IVIFS set, IVIFS basis, and IVIFS subspace. Some examples and theorems regarding these concepts are presented.

Definition 5.1 [4] Union of two interval-valued intuitionistic fuzzy soft sets (f, A) and (g, B) over (U, E) is an interval-valued intuitionistic fuzzy soft set (h, C) where $C = A \cup B$ and $\forall e \in C$,

$$h(e) = \begin{cases} f(e), & \text{if } e \in A - B \\ g(e), & \text{if } e \in B - A \\ f(e) \cup g(e), & \text{if } e \in A \cap B \end{cases}$$

and is written as $(f, A) \cup (g, B) = (h, C)$.

Definition 5.2 [4] Intersection of two interval-valued intuitionistic fuzzy soft sets (f, A) and (g, B) over (U, E) is an interval-valued intuitionistic fuzzy soft set (h, C) where $C = A \cup B$ and $\forall e \in C$,

$$h(e) = \begin{cases} f(e), & \text{if } e \in A - B \\ g(e), & \text{if } e \in B - A \\ f(e) \cap g(e), & \text{if } e \in A \cap B \end{cases}$$

and is written as $(f, A) \cup (g, B) = (h, C)$.

Definition 5.3 [4] For two interval-valued intuitionistic fuzzy soft sets (f, A) and (g, B) over (U, E). Then, we say that (f, A) is an interval-valued intuitionistic fuzzy soft subset of (g, B), if

(i) $A \subseteq B$
(ii) for all $e \in A, F(e) \subseteq G(e)$.

Definition 5.4 [9, 12] Let (f_A, E) be a fuzzy soft set on (U, E) and τ_f be the collection of fuzzy soft subsets of (f_A, E), and then, τ is said to be fuzzy soft topology on (f_A, E) if the following conditions hold:

[O_1]. $\Phi, (f_A, E) \in \tau_f$ (where Φ has been defined in [10])
[O_2]. $\{(f_A^k, E)|k \in K\} \subset \tau_f \Rightarrow \cup_{k \in K}(f_A^k, E) \in \tau_f$
[O_3]. $(g_A, E), (h_A, E) \subset \tau_f \Rightarrow (g_A, E) \cap (h_A, E) \in \tau_f$

The triplet (f_A, E, τ_f) is called a fuzzy soft topology over (f_A, E).

Let U be the initial universe, E be the set of parameters, $P(U)$ be the set of all subsets of U, IVIFS set (U) be the set of all interval-valued intuitionistic fuzzy sets in U, and IVIFS $(U; E)$ be the family of all intuitionistic fuzzy soft sets over U via parameters in Simsekler and Yuksel [11] introduced fuzzy soft topology over a fuzzy soft set with a fixed parameter set $A \subseteq E$. Zhaowen et al. [14] defined the topological structure of intuitionistic fuzzy soft sets taking the whole parameter set E. In this section, we introduced the concept of interval-valued intuitionistic fuzzy soft topological spaces with a fixed parameter set $A \subseteq E$, which is the extension of fuzzy soft topological spaces introduced by Simsekler and Yuksel [11] as well as intuitionistic fuzzy soft topological spaces introduced by Zhaowen et al. [14].

Definition 5.5 Let (ξ_A, E) be an element of IVIFS $(U; E)$ and $P(\xi_A, E)$ be the collection of all IVIFS subsets of (ξ_A, E). A subfamily τ of $P(\xi_A, E)$ is called an *interval-valued intuitionistic fuzzy soft topology* (*in short IVIFS topology*) on (ξ_A, E) if the following axioms are satisfied:

$[O_1]$. $(\phi_{\xi_A}, E), (\xi_A, E) \in \tau$,

$[O_2]$. $\left\{ (f_A^k, E) | k \in K \right\} \subseteq \tau \Rightarrow \cup_{k \in K} (f_A^k, E) \in \tau$

$[O_3]$. If $(f_A, E), (g_A, E) \in \tau$, then $(f_A, E) \cap (g_A, E) \in \tau$

Then, the pair $((\xi_A, E), \tau)$ is called an interval-valued intuitionistic fuzzy soft topological space (*IVIFS topological space for short*). The members of τ are called *τ-open IVIFS sets* (or simply *open sets*). (where $\phi_{\xi_A} : A \to$ IVIFS(U) is defined as $\phi_{\xi_A}(e) = \{\langle x, [0,0], [1,1] \rangle : x \in U\}, \forall e \in A$.)

Example 5.6 Let $U = \{u^1, u^2, u^3\}$, $E = \{e_1, e_2, e_3, e_4\}$, $A = \{e_1, e_2, e_3\}$ and

$$(\xi_A, E) = \left\{ e_1 = \{ u^1_{([1,1],[0,0])}, u^2_{([0.7,0.8],[0,0])}, u^3_{([1,1],[0,0])} \}, \right.$$
$$e_2 = \{ u^1_{([0.4,0.5],[0,0])}, u^2_{([1,1],[0,0])}, u^3_{([0.4,0.5],[0.2,0.3])} \},$$
$$\left. e_3 = \{ u^1_{([0,0],[1,1])}, u^2_{([0,0],[1,1])}, u^3_{([0.4,0.5],[0.1,0.2])} \} \right\}$$

$$(\phi_{\xi_A}, E) = \left\{ e_1 = \{ u^1_{([0,0],[1,1])}, u^2_{([0,0],[1,1])}, u^3_{([0,0],[1,1])} \}, \right.$$
$$e_2 = \{ u^1_{([0,0],[1,1])}, u^2_{([0,0],[1,1])}, u^3_{([0,0],[1,1])} \},$$
$$\left. e_3 = \{ u^1_{([0,0],[1,1])}, u^2_{([0,0],[1,1])}, u^3_{([0,0],[1,1])} \} \right\}$$

$$(f_A^1, E) = \left\{ e_1 = \{ u^1_{([0.5,0.6],[0.2,0.3])}, u^2_{([0.4,0.5],[0,0.1])}, u^3_{([1,1],[0,0])} \}, \right.$$
$$e_2 = \{ u^1_{([0.4,0.5],[0.2,0.3])}, u^2_{([0.4,0.5],[0,0])}, u^3_{([0,0],[1,1])} \},$$
$$\left. e_3 = \{ u^1_{([0,0],[1,1])}, u^2_{([0,0],[1,1])}, u^3_{([0,0],[1,1])} \} \right\}$$

$$(f_A^2, E) = \left\{ e_1 = \{ u^1_{([0.3,0.4],[0.1,0.2])}, u^2_{([0.6,0.7],[0.2,0.3])}, u^3_{([1,1],[0,0])} \}, \right.$$
$$e_2 = \{ u^1_{([0.2,0.3],[0,0.1])}, u^2_{([1,1],[0,0])}, u^3_{([0,0],[1,1])} \},$$
$$\left. e_3 = \{ u^1_{([0,0],[1,1])}, u^2_{([0,0],[1,1])}, u^3_{([0,0],[1,1])} \} \right\}$$

$$\left(f_A^3, E\right) = \left(f_A^1, E\right) \cap \left(f_A^2, E\right) = \Big\{ e_1 = \{u^1_{([0.3,0.4],[0.2,0.3])}, u^2_{([0.4,0.5],[0.2,0.3])}, u^3_{([1,1],[0,0])}\},$$

$$e_2 = \{u^1_{([0.2,0.3],[0.2,0.3])}, u^2_{([0.4,0.5],[0,0])}, u^3_{([0,0],[1,1])}\},$$

$$e_3 = \{u^1_{([0,0],[1,1])}, u^2_{([0,0],[1,1])}, u^3_{([0,0],[1,1])}\}\Big\}$$

$$\left(f_A^4, E\right) = \left(f_A^1, E\right) \cup \left(f_A^2, E\right) = \Big\{ e_1 = \{u^1_{([0.5,0.6],[0.1,0.2])}, u^2_{([0.6,0.7],[0,0.1])}, u^3_{([1,1],[0,0])}\},$$

$$e_2 = \{u^1_{([0.4,0.5],[0,0.1])}, u^2_{([1,1],[0,0])}, u^3_{([0,0],[1,1])}\},$$

$$e_3 = \{u^1_{([0,0],[1,1])}, u^2_{([0,0],[1,1])}, u^3_{([0,0],[1,1])}\}\Big\}$$

Then, we observe that the subfamily $\tau_1 = \big\{ (\phi_{\xi_A}, E), (\xi_A, E), (f_A^1, E), (f_A^2, E),$ $(f_A^3, E), (f_A^4, E)\big\}$ of $P(\xi_A, E)$ is a IVIFS topology on (ξ_A, E) since it satisfies the necessary three axioms $[O_1]$, $[O_2]$ and $[O_3]$ and $((\xi_A, E), \tau)$ is an *interval-valued intuitionistic fuzzy soft topological space*. But the subfamily $\tau_2 = \big\{ (\phi_{\xi_A}, E),$ $(\xi_A, E), (f_A^1, E), (f_A^2, E)\big\}$ of $P(\xi_A, E)$ is not an *IVIFS topology* on (ξ_A, E) since the union $(f_A^1, E) \cup (f_A^2, E) = (f_A^4, E)$ which does not belong to τ_2.

Definition 5.7 As every IVIFS topology on (ξ_A, E) must contain the sets (ϕ_{ξ_A}, E) and (ξ_A, E), so the family $\vartheta = \big\{ (\phi_{\xi_A}, E), (\xi_A, E)\big\}$ forms an IVIFS topology on (ξ_A, E). This topology is called *indiscrete IVIFS topology* and the pair $((\xi_A, E), \vartheta)$ is called an indiscrete interval-valued intuitionistic fuzzy soft topological space (or simply *indiscrete IVIFS topological space*).

Definition 5.8 Let D denote family of all IVIFS subsets of (ξ_A, E). Then, we observe that D satisfies all the axioms for topology on (ξ_A, E). This topology is called *discrete IVIFS topology* and the pair $((\xi_A, E), D)$ is called a discrete interval-valued intuitionistic fuzzy soft topological space (or simply *discrete IVIFS topological space*).

Theorem 5.9 *Let $\{\tau_i : i \in I\}$ be any collection of IVIFS topology on (ξ_A, E). Then, their intersection $\cap_{i \in I} \tau_i$ is also a topology on (ξ_A, E).*

Proof

$[O_1]$. Since (ϕ_{ξ_A}, E), $(\xi_A, E) \in \tau_i$, for each $i \in I$, hence (ϕ_{ξ_A}, E), $(\xi_A, E) \in \cap_{i \in I} \tau_i$

$[O_2]$. Let $\{(f_A^k, E) | k \in K\}$ be an arbitrary family of interval-valued intuitionistic fuzzy soft sets where $(f_A^k, E) \in \cap_{i \in I} \tau_i$ for each $k \in K$. Then, for each $i \in I$, $(f_A^k, E) \in \tau_i$ for $k \in K$, and since for each $i \in I$, τ_i is an topology; therefore, $\cup_{k \in K}(f_A^k, E) \in \tau_i$, for each $i \in I$. Hence, $\cup_{k \in K}(f_A^k, E) \in \cap_{i \in I} \tau_i$

$[O_3]$. Let (f_A, E) and $(g_A, E) \in \cap_{i \in I} \tau_i$, and then (f_A, E) and $(g_A, E) \in \tau_i$, for each $i \in I$, and since τ_i for each $i \in I$ is an topology, therefore $(f_A, E) \cap (g_A, E) \in \tau_i$ for each $i \in I$, Hence, $(f_A, E) \cap (g_A, E) \in \cap_{i \in I} \tau_i$.

Thus, $\cap_{i \in I} \tau_i$ satisfies all the axioms of topology. Hence, $\cap_{i \in I} \tau_i$ forms a topology.

But union of topologies need not be a topology; we can show this with following example.   □

Remark 5.10 The union of two IVIFS topologies may not be a IVIFS topology. If we consider the Example 5.6, then the subfamilies $\tau_3 = \{(\phi_{\xi_A}, E), (\xi_A, E), (f_A^1, E)\}$ and $\tau_4 = \{(\phi_{\xi_A}, E), (\xi_A, E), (f_A^2, E)\}$ are the topologies in (ξ_A, E), but their union $\tau_3 \cup \tau_4 = \{(\phi_{\xi_A}, E), (\xi_A, E), (f_A^1, E), (f_A^2, E)\} = \tau_2$ which is not a topology on (ξ_A, E).

Definition 5.11 Let $((\xi_A, E), \tau)$ be an *IVIFS topological* space over (ξ_A, E). An IVIFS subset (f_A, E) of (ξ_A, E) is called interval-valued intuitionistic fuzzy soft closed (in short IVIFS closed) if its complement $(f_A, E)^c$ is a member of τ.

Example 5.12 Let us consider Example 5.6, and then, the IVIFS closed sets in $((\xi_A, E), \tau_1)$ are:

$$(\phi_{\xi_A}, E)^c = (U, E) = \Big\{ e_1 = \{u^1_{([1,1],[0,0])}, u^2_{([1,1],[0,0])}, u^3_{([1,1],[0,0])}\},$$

$$e_2 = \{u^1_{([1,1],[0,0])}, u^2_{([1,1],[0,0])}, u^3_{([1,1],[0,0])}\},$$

$$e_3 = \{u^1_{([1,1],[0,0])}, u^2_{([1,1],[0,0])}, u^3_{([1,1],[0,0])}\}\Big\}$$

$$(\xi_A, E)^c = \Big\{ e_1 = \{u^1_{([0,0],[1,1])}, u^2_{([0,0],[0.7,0.8])}, u^3_{([0,0],[1,1])}\},$$

$$e_2 = \{u^1_{([0,0],[0.4,0.5])}, u^2_{([0,0],[1,1])}, u^3_{([0.2,0.3],[0.4,0.5])}\},$$

$$e_3 = \{u^1_{([1,1],[0,0])}, u^2_{([1,1],[0,0])}, u^3_{([0.1,0.2],[0.4,0.5])}\}\Big\}$$

$$(f_A^1, E)^c = \Big\{ e_1 = \{u^1_{([0.2,0.3],[0.5,0.6])}, u^2_{([0,0.1],[0.4,0.5])}, u^3_{([0,0],[1,1])}\},$$

$$e_2 = \{u^1_{([0.2,0.3],[0.4,0.5])}, u^2_{([0,0],[0.4,0.5])}, u^3_{([1,1],[0,0])}\},$$

$$e_3 = \{u^1_{([1,1],[0,0])}, u^2_{([1,1],[0,0])}, u^3_{([1,1],[0,0])}\}\Big\}$$

$$(f_A^2, E)^c = \Big\{ e_1 = \{u^1_{([0.1,0.2],[0.3,0.4])}, u^2_{([0.2,0.3],[0.6,0.7])}, u^3_{([0,0],[1,1])}\},$$

$$e_2 = \{u^1_{([0,0.1],[0.2,0.3])}, u^2_{([0,0],[1,1])}, u^3_{([1,1],[0,0])}\},$$

$$e_3 = \{u^1_{([1,1],[0,0])}, u^2_{([1,1],[0,0])}, u^3_{([1,1],[0,0])}\}\Big\}$$

$$(f_A^3, E)^c = \Big\{ e_1 = \{u^1_{([0.2,0.3],[0.3,0.4])}, u^2_{([0.2,0.3],[0.4,0.5])}, u^3_{([0,0],[1,1])}\},$$

$$e_2 = \{u^1_{([0.2,0.3],[0.2,0.3])}, u^2_{([0,0],[0.4,0.5])}, u^3_{([1,1],[0,0])}\},$$

$$e_3 = \{u^1_{([1,1],[0,0])}, u^2_{([1,1],[0,0])}, u^3_{([1,1],[0,0])}\}\Big\}$$

$$(f_A^4, E)^c = \Big\{ e_1 = \{ u^1_{([0.1,0.2],[0.5,0.6])}, u^2_{([0,0.1],[0.6,0.7])}, u^3_{([0,0],[1,1])} \},$$

$$e_2 = \{ u^1_{([0,0.1],[0.4,0.5])}, u^2_{([0,0],[1,1])}, u^3_{([1,1],[0,0])} \},$$

$$e_3 = \{ u^1_{([1,1],[0,0])}, u^2_{([1,1],[0,0])}, u^3_{([1,1],[0,0])} \} \Big\}$$

which are the intuitionistic fuzzy soft closed sets in $((\xi_A, E), \tau)$.

Theorem 5.13 *Let* $((\xi_A, E), \tau)$ *be an interval-valued intuitionistic fuzzy soft topological space over* (ξ_A, E). *Then,*

1. $(\phi_{\xi_A}, E)^c$ *and* $(\xi_A, E)^c$ *are interval-valued intuitionistic fuzzy soft closed sets,*
2. *The arbitrary intersection of interval-valued intuitionistic fuzzy soft closed sets is interval-valued intuitionistic fuzzy soft closed,*
3. *The union of two interval-valued intuitionistic fuzzy soft closed sets is an interval-valued intuitionistic fuzzy soft closed set.*

Proof

1. Since $(\phi_{\xi_A}, E), (\xi_A, E) \in \tau, (\phi_{\xi_A}, E)^c$ and $(\xi_A, E)^c$ are closed.
2. Let $\{ (f_A^k, E) | k \in K \}$ be an arbitrary family of IVIFS closed sets in $((\xi_A, E), \tau)$ and let $(f_A, E) = \cap_{k \in K} (f_A^k, E)$.
   Now, since $(f_A, E)^c = (\cap_{k \in K} (f_A^k, E))^c = \cup_{k \in K} (f_A^k, E)^c$ and $(f_A^k, E)^c \in \tau$, for each $k \in K$, $\cup_{k \in K} (f_A^k, E)^c \in \tau$. Hence, $(f_A, E)^c \in \tau$. Thus, (f_A, E) is IVIFS closed set.
3. Let $\{ (f_A^i, E) | i = 1, 2, 3, \ldots, n \}$ be a finite family of IVIFS closed sets in $((\xi_A, E), \tau)$ and let $(g_A, E) = \cup_{i=1}^n (f_A^i, E)$.
   Now, since $(g_A, E)^C = (\cup_{i=1}^n (f_A^i, E))^c = \cap_{i=1}^n (f_A^i, E)^c$ and $(f_A^k, E)^c \in \tau$.
   So $\cap_{i=1}^n (f_A^i, E)^c \in \tau$. Hence, $(g_A, E)^c \in \tau$. Thus, (g_A, E) is an IVIFS closed set. $\square$

Remark 5.14 The intersection of an arbitrary family of IVIFS open set may not be an IVIFS open, and the union of an arbitrary family of IVIFS closed set may not be an IVIFS closed. Let us consider $U = \{ u^1, u^2, u^3 \}$, $E = \{ e_1, e_2, e_3, e_4 \}$, $A = \{ e_1, e_2, e_3 \}$ and let

$$(\xi_A, E) = \Big\{ e_1 = \{ u^1_{([1,1],[0,0])}, u^2_{([1,1],[0,0])}, u^3_{([0,0],[1,1])} \},$$

$$e_2 = \{ u^1_{([1,1],[0,0])}, u^2_{([0,0],[1,1])}, u^3_{([0,0],[1,1])} \},$$

$$e_3 = \{ u^1_{([0,0],[1,1])}, u^2_{([0,0],[1,1])}, u^3_{([0,0],[1,1])} \} \Big\}$$

$$(\phi_{\xi_A}, E) = \Big\{ e_1 = \{u^1_{([0,0],[1,1])}, u^2_{([0,0],[1,1])}, u^3_{([0,0],[1,1])}\},$$
$$e_2 = \{u^1_{([0,0],[1,1])}, u^2_{([0,0],[1,1])}, u^3_{([0,0],[1,1])}\},$$
$$e_3 = \{u^1_{([0,0],[1,1])}, u^2_{([0,0],[1,1])}, u^3_{([0,0],[1,1])}\} \Big\}$$

For each $n \in N$, we define

$$(f_A^n, E) = \Big\{ e_1 = \{u^1_{([\frac{1}{4n},\frac{1}{2n}],[\frac{1}{5}-\frac{1}{2n},\frac{1}{2}-\frac{1}{3n}])}, u^2_{([1,1],[0,0])}, u^3_{([0,0],[1,1])}\},$$
$$e_2 = \{u^1_{([\frac{1}{3n},\frac{1}{2n}],[\frac{1}{3}-\frac{1}{3n},\frac{1}{3}-\frac{1}{4n}])}, u^2_{([0,0],[1,1])}, u^3_{([0,0],[1,1])}\},$$
$$e_3 = \{u^1_{([0,0],[1,1])}, u^2_{([0,0],[1,1])}, u^3_{([0,0],[1,1])}\} \Big\}$$

Let us consider the subfamily τ of $P(\xi_A, E)$, such that $(\phi_{\xi_A}, E), (\xi_A, E) \in \tau$ and

$$(f_A^n, E) \in \tau \text{ for } n = 1, 2, 3, \ldots$$

Then, we observe that τ is a IVIFS topology on (ξ_A, E).
But

$$\bigcap_{n=1}^{\infty} (f_A^n, E) = \Big\{ e_1 = \{u^1_{([0,0],[0.2,0.5])}, u^2_{([1,1],[0,0])}, u^3_{([0,0],[1,1])}\},$$
$$e_2 = \{u^1_{([0,0],[0.33,0.5])}, u^2_{([0,0],[1,1])}, u^3_{([0,0],[1,1])}\},$$
$$e_3 = \{u^1_{([0,0],[1,1])}, u^2_{([0,0],[1,1])}, u^3_{([0,0],[1,1])}\} \Big\} \notin \tau.$$

The IVIFS closed sets in the IVIFS topological space $((\xi_A, E), \tau)$ are:

$$(\phi_{\xi_A}, E)^c, (\xi_A, E)^c, (f_A^n, E)^c \quad (\text{for } n = 1, 2, 3, \ldots)$$

But

$$\bigcup_{n=1}^{\infty} (f_A^n, E)^C = \Big\{ e_1 = \{u^1_{([0.2,0.5],[0,0])}, u^2_{([0,0],[1,1])}, u^3_{([1,1],[0,0])}\},$$
$$e_2 = \{u^1_{([0.33,0.5],[0,0])}, u^2_{([1,1],[0,0])}, u^3_{([1,1],[0,0])}\},$$
$$e_3 = \{u^1_{([1,1],[0,0])}, u^2_{([1,1],[0,0])}, u^3_{([1,1],[0,0])}\} \Big\}$$

is not an IVIFS closed set in IVIFS topological space $((\xi_A, E), \tau)$, since $\left(\bigcup_{n=1}^{\infty} (f_A^n, E)^c \right)^c \notin \tau$.

Definition 5.15 Let $((\xi_A, E), \tau_1)$ and $((\xi_A, E), \tau_2)$ be two IVIFS topological spaces. If each $(f_A, E) \in \tau_1 \Rightarrow (f_A, E) \in \tau_2$, then τ_2 is called interval-valued intuitionistic fuzzy soft finer topology than τ_1 and τ_1 is called interval-valued intuitionistic fuzzy soft coarser topology than τ_2.

Example 5.16 If we consider the topologies $\tau_1 = \{(\phi_{\xi_A}, E), (\xi_A, E), (f_A^1, E),$ $(f_A^2, E), (f_A^3, E), (f_A^4, E)\}$ as in the Example 5.6 and $\tau_5 = \{(\phi_{\xi_A}, E), (\xi_A, E),$ $(f_A^3, E)\}$ on (ξ_A, E).

Then, τ_1 is interval-valued intuitionistic fuzzy soft finer topology than τ_5 and τ_5 is interval-valued intuitionistic fuzzy soft coarser topology than τ_1.

Definition 5.17 Let $((\xi_A, E), \tau)$ be an IVIFS topological space on (ξ_A, E) and B be a subfamily of τ. If every element of τ can be expressed as the arbitrary interval-valued intuitionistic fuzzy soft union of some element of B, then B is called an interval-valued intuitionistic fuzzy soft basis for the interval-valued intuitionistic fuzzy soft topology τ.

Example 5.18 In the Example 5.6, for the topology $\tau_1 = \{(\phi_{\xi_A}, E), (\xi_A, E),$ $(f_A^1, E), (f_A^2, E), (f_A^3, E), (f_A^4, E)\}$, the subfamily $B = \{(\phi_{\xi_A}, E), (\xi_A, E), (f_A^1, E),$ $(f_A^2, E), (f_A^3, E)\}$ of $P(\xi_A, E)$ is a basis for the topology τ_1.

5.1 Neighbourhoods and Neighbourhood Systems

Definition 5.19 Let τ be the IVIFS topology on $(\xi_A, E) \in IFS(U; E)$ and (f_A, E) be an IVIFS set in $P(\xi_A, E)$. A IVIFS set (f_A, E) in $P(\xi_A, E)$ is a neighbourhood of a IVIFS set (g_A, E) if and only if there exists an τ-open IVIFS set (h_A, E), i.e. $(h_A, E) \in \tau$ such that $(g_A, E) \subseteq (h_A, E) \subseteq (f_A, E)$.

Example 5.20 In an IVIFS topology

$$\tau = \left\{ (\phi_{\xi_A}, E), (\xi_A, E), \left\{ e_1 = \{u^1_{([0.4,0.5],[0.4,0.5])}, u^2_{([0.3,0.4],[0.5,0.6])}, u^3_{([0.4,0.5],[0.1,0.2])}\} \right\} \right\}$$

where

$$(\xi_A, E) = \left\{ e_1 = \{u^1_{([1,1],[0,0])}, u^2_{([0.7,0.8],[0,0])}, u^3_{([1,1],[0,0])}\}, \right.$$
$$e_2 = \{u^1_{([0.4,0.5],[0,0])}, u^2_{([1,1],[0,0])}, u^3_{([0.4,0.5],[0.2,0.3])}\},$$
$$\left. e_3 = \{u^1_{([0,0],[1,1])}, u^2_{([0,0],[1,1])}, u^3_{([0.4,0.5],[0.1,0.2])}\} \right\}$$

and

$$(\phi_{\xi_A}, E) = \left\{ e_1 = \{u^1_{([0,0],[1,1])}, u^2_{([0,0],[1,1])}, u^3_{([0,0],[1,1])}\}, \right.$$
$$e_2 = \{u^1_{([0,0],[1,1])}, u^2_{([0,0],[1,1])}, u^3_{([0,0],[1,1])}\},$$
$$\left. e_3 = \{u^1_{([0,0],[1,1])}, u^2_{([0,0],[1,1])}, u^3_{([0,0],[1,1])}\} \right\}$$

The IVIFS set $(f_A, E) = \left\{ e_1 = \{ u^1_{([0.5,0.6],[0.2,0.3])}, u^2_{([0.3,0.4],[0.5,0.6])}, u^3_{([0.4,0.5],[0.0,0.1])} \} \right\}$
is a neighbourhood of the IVIFS set $(g_A, E) = \left\{ e_1 = \{ u^1_{([0.3,0.4],[0.4,0.5])}, u_{([0.1,0.2],} \right.$
$\left. [0.6,0.7])}^2, u^3_{([0.4,0.5],[0.3,0.4])} \} \right\}$, because there exists an τ-open IVIFS set $(h_A, E) = $
$\left\{ e_1 = \{ u^1_{([0.4,0.5],[0.4,0.5])}, u^2_{([0.3,0.4],[0.5,0.6])}, u^3_{([0.4,0.5],[0.1,0.2])} \} \right\} \in \tau$ such that $(g_A, E) \subseteq$
$(h_A, E) \subseteq (f_A, E)$.

Theorem 5.20 *A IVIFS set (f_A, E) in $P(\xi_A, E)$ is an open IVIFS set if and only if (f_A, E) is a neighbourhood of each IVIFS set (g_A, E) contained in (f_A, E).*

Proof Let (f_A, E) be an open IVIFS set and (g_A, E) be any IVIFS set contained in (f_A, E). Since we have $(g_A, E) \subseteq (f_A, E) \subseteq (f_A, E)$, it follows that (f_A, E) is a neighbourhood of (g_A, E).

Conversely, let (f_A, E) be a neighbourhood for every IVIFS sets contained it. Since $(f_A, E) \subseteq (f_A, E)$, there exist an open IVIFS set (h_A, E) such that $(f_A, E) \subseteq (h_A, E) \subseteq (f_A, E)$. Hence, $(f_A, E) = (h_A, E)$ and (f_A, E) is open. $\square$

Definition 5.21 Let $((\xi_A, E), \tau)$ be an interval-valued intuitionistic fuzzy soft topological space on (ξ_A, E) and (f_A, E) be a IVIFS set in $P(\xi_A, E)$. The family of all neighbourhoods of (f_A, E) is called the neighbourhood system of (f_A, E) up to topology and is denoted by $N_{(f_A, E)}$.

Theorem 5.22 *Let $((\xi_A, E), \tau)$ be an interval-valued intuitionistic fuzzy soft topological space. If $N_{(f_A, E)}$ is the neighbourhood system of an interval-valued intuitionistic fuzzy soft set (f_A, E), then,*

1. *$N_{(f_A, E)}$ is non-empty and (f_A, E) belongs to the each member of $N_{(f_A, E)}$.*
2. *The intersection of any two members of $N_{(f_A, E)}$ belongs to $N_{(f_A, E)}$.*
3. *Each interval-valued intuitionistic fuzzy soft set which contains a member of $N_{(f_A, E)}$ belongs to $N_{(f_A, E)}$.*

Proof

1. If $(h_A, E) \in N_{(f_A, E)}$, then there exist an open set $(g_A, E) \in \tau$ such that $(f_A, E) \subseteq (g_A, E) \subseteq (h_A, E)$; hence, $(f_A, E) \subseteq (h_A, E)$. Note $(\xi_A, E) \in N_{(f_A, E)}$, and since (ξ_A, E) is an open set containing (f_A, E), $N_{(f_A, E)}$ is non-empty.
2. Let (g_A, E) and (h_A, E) are two neighbourhoods of (f_A, E), so there exist two open sets (g^*_A, E), (h^*_A, E) such that $(f_A, E) \subseteq (g^*_A, E) \subseteq (g_A, E)$ and $(f_A, E) \subseteq (h^*_A, E) \subseteq (h_A, E)$. Hence, $(f_A, E) \subseteq (g^*_A, E) \cap (h^*_A, E) \subseteq (g_A, E) \cap (h_A, E)$ and $(g^*_A, E) \cap (h^*_A, E)$ are open. Thus, $(g_A, E) \cap (h_A, E)$ is a neighbourhoods of (f_A, E).
3. Let (g_A, E) is a neighbourhood of (f_A, E) and $(g_A, E) \subseteq (h_A, E)$, so there exist an open set (g^*_A, E), such that $(f_A, E) \subseteq (g^*_A, E) \subseteq (g_A, E)$. By hypothesis $(g_A, E) \subseteq (h_A, E)$, so $(f_A, E) \subseteq \left(g^*_A, E \right) \subseteq (g_A, E) \subseteq (h_A, E)$, which implies that $(f_A, E) \subseteq (g^*_A, E) \subseteq (h_A, E)$ and hence (h_A, E) is a neighbourhood of (f_A, E). $\square$

5.2 Interior, Exterior, and Closure

Definition 5.23 Let $((\xi_A, E), \tau)$ be an interval-valued intuitionistic fuzzy soft topological space on (ξ_A, E) and (f_A, E), (g_A, E) be IVIFS sets in $P(\xi_A, E)$ such that $(g_A, E) \subseteq (f_A, E)$. Then, (g_A, E) is called an interior IVIFS set of (f_A, E) if and only if (f_A, E) is a neighbourhood of (g_A, E).

Definition 5.24 Let $((\xi_A, E), \tau)$ be an interval-valued intuitionistic fuzzy soft topological space on (ξ_A, E) and (f_A, E) be an IVIFS set in $P(\xi_A, E)$. Then, the union of all interior IVIFS set of (f_A, E) is called the interior of (f_A, E) and is denoted by $\text{int}(f_A, E)$ and defined by

$$\text{int}(f_A, E) = \cup\{(g_A, E) | (f_A, E) \text{ is a neighbourhood of } (g_A, E)\}.$$

Or equivalently

$$\text{int}(f_A, E) = \cup\{(g_A, E) | (g_A, E) \text{ is an IVIFS-open set contained in } (f_A, E)\}.$$

Example 5.25 Let us consider the IVIFS topology $\tau_1 = \{(\phi_{\xi_A}, E), (\xi_A, E), (f_A^1, E), (f_A^2, E), (f_A^3, E), (f_A^4, E)\}$ as in the Example 5.6 and let

$$(f_A, E) = \Big\{ e_1 = \{u^1_{([0.1,0.5],[0.1,0.2])}, u^2_{([0.6,0.7],[0.2,0.3])}, u^3_{([1,1],[0,0])}\},$$
$$e_2 = \{u^1_{([0.3,0.4],[0,0.1])}, u^2_{([1,1],[0,0])}, u^3_{([0,0],[1,1])}\},$$
$$e_3 = \{u^1_{([0,0],[1,1])}, u^2_{([0,0],[1,1])}, u^3_{([0,0],[1,1])}\} \Big\}$$

be an IVIFS set, and then,

$$\text{int}(f_A, E) = \cup\{(g_A, E) | (g_A, E) \text{ is an IVIFS-open set contained in } (f_A, E)\}$$
$$= (f_A^2, E) \cup (f_A^3, E)$$
$$= (f_A^2, E)$$
$$= \Big\{ e_1 = \{u^1_{([0.3,0.4],[0.1,0.2])}, u^2_{([0.6,0.7],[0.2,0.3])}, u^3_{([1,1],[0,0])}\},$$
$$e_2 = \{u^1_{([0.2,0.3],[0,0.1])}, u^2_{([1,1],[0,0])}, u^3_{([0,0],[1,1])}\},$$
$$e_3 = \{u^1_{([0,0],[1,1])}, u^2_{([0,0],[1,1])}, u^3_{([0,0],[1,1])}\} \Big\}.$$

Since $(f_A^2, E) \subseteq (f_A, E)$ and $(f_A^3, E) \subseteq (f_A, E)$.

Theorem 5.26 *Let $((\xi_A, E), \tau)$ be an interval-valued intuitionistic fuzzy soft topological space on (ξ_A, E) and (f_A, E) be an IVIFS set in $P(\xi_A, E)$. Then,*

1. $\text{int}(f_A, E)$ *is an open and* $\text{int}(f_A, E)$ *is the largest open IVIFS set contained in* (f_A, E).
2. *The IVIFS set* (f_A, E) *is open if and only if* $(f_A, E) = \text{int}(f_A, E)$.

Proof

1. Since $\text{int}(f_A, E) = \cup\{(g_A, E) | (f_A, E) \text{ is a neighbourhood of } (g_A, E)\}$, we have that (f_A, E) is itself an interior IVIFS set of (f_A, E). Then, there exists an open IVIFS set (h_A, E) such that $\text{int}(f_A, E) \subseteq (h_A, E) \subseteq (f_A, E)$. But (h_A, E) is an interior IVIFS set of (f_A, E), hence $(h_A, E) \subseteq \text{int}(f_A, E)$. Hence, $(h_A, E) = \text{int}(f_A, E)$. Thus, $\text{int}(f_A, E)$ is open and $\text{int}(f_A, E)$ is the largest open IVIFS set contained in (f_A, E).

2. Let (f_A, E) be an open IVIFS set. Since $\text{int}(f_A, E)$ is an interior IVIFS set of (f_A, E), we have $(f_A, E) = \text{int}(f_A, E)$. Conversely, if $(f_A, E) = \text{int}(f_A, E)$, then (f_A, E) is obviously open. $\qquad\square$

Proposition 5.27 *For any two IVIFS sets (f_A, E) and (g_A, E) in an interval-valued intuitionistic fuzzy soft topological space $((\xi_A, E), \tau)$ on $P(\xi_A, E)$, then*

(i) $(g_A, E) \subseteq (f_A, E) \Rightarrow \text{int}(g_A, E) \subseteq \text{int}(f_A, E)$

(ii) $\text{int}(\phi_{\xi_A}, E) = (\phi_{\xi_A}, E)$ *and* $\text{int}(\xi_A, E) = (\xi_A, E)$

(iii) $\text{int}(\text{int}(f_A, E)) = \text{int}(f_A, E)$

(iv) $\text{int}((g_A, E) \cap (f_A, E)) = \text{int}(g_A, E) \cap \text{int}(f_A, E)$

(v) $\text{int}((g_A, E) \cup (f_A, E)) \supseteq \text{int}(g_A, E) \cup \text{int}(f_A, E)$

Proof Since $(g_A, E) \subseteq (f_A, E)$, implies all the open set contained in (g_A, E) also contained in (f_A, E). Therefore,

$$\{(g_A^*, E) | (g_A^*, E) \text{ is an IVIFS-open set contained in } (g_A, E)\}$$
$$\subseteq \{(f_A^*, E) | (f_A^*, E) \text{ is an IVIFS-open set contained in } (f_A, E)\}$$

which implies

(i) $\cup\{(g_A^*, E) | (g_A^*, E) \text{ is an IVIFS-open set contained in } (g_A, E)\} \subseteq \cup\{(f_A^*, E) | (f_A^*, E) \text{ is an IVIFS-open set contained in } (f_A, E)\} \Rightarrow \text{int}(g_A, E) \subseteq \text{int}(f_A, E)$

(ii) $\text{int}(\text{int}(f_A, E)) = \cup\{(g_A, E) | (g_A, E) \text{ is an IVIFS-open set contained in } \text{int}(f_A, E)\}$, and since $\text{int}(f_A, E)$ is the largest open IVIFS set contained in $\text{int}(f_A, E)$, therefore $\text{int}(\text{int}(f_A, E)) = \text{int}(f_A, E)$.

(iii) Since $\text{int}((g_A, E)) \subseteq (g_A, E)$ and $\text{int}((f_A, E)) \subseteq (f_A, E)$

$$\Rightarrow \text{int}(g_A, E) \cap \text{int}(f_A, E) \subseteq (g_A, E) \cap (f_A, E)$$
$$\Rightarrow \text{int}(g_A, E) \cap \text{int}(f_A, E) \subseteq \text{int}((g_A, E) \cap (f_A, E)) \tag{5.1}$$

Again since $(g_A, E) \cap (f_A, E) \subseteq (g_A, E)$ and $(g_A, E) \cap (f_A, E) \subseteq (f_A, E)$

$$\Rightarrow \text{int}((g_A, E) \cap (f_A, E)) \subseteq \text{int}(g_A, E) \text{ and } \text{int}((g_A, E) \cap (f_A, E)) \subseteq \text{int}(f_A, E)$$
$$\Rightarrow \text{int}((g_A, E) \cap (f_A, E)) \subseteq \text{int}(g_A, E) \cap \text{int}(f_A, E) \tag{5.2}$$

Using (5.1) and (5.2), we get $\text{int}((g_A, E) \cap (f_A, E)) = \text{int}(g_A, E) \cap \text{int}(f_A, E)$

(iv) Since $(g_A, E) \subseteq (g_A, E) \cup (f_A, E)$ and $(f_A, E) \subseteq (g_A, E) \cup (f_A, E)$

$$\Rightarrow \text{int}(g_A, E) \subseteq \text{int}((g_A, E) \cup (f_A, E)) \text{ and int}(f_A, E) \subseteq \text{int}((g_A, E) \cup (f_A, E))$$
$$\Rightarrow \text{int}(g_A, E) \cup \text{int}(f_A, E) \subseteq \text{int}((g_A, E) \cup (f_A, E))$$

$\square$

Definition 5.28 Let $((\xi_A, E), \tau)$ be an interval-valued intuitionistic fuzzy soft topological space on (ξ_A, E) and let (f_A, E) and (g_A, E) be two IVIFS sets in $P(\xi_A, E)$. Then, (g_A, E) is called an exterior IVIFS set of (f_A, E) if and only if (g_A, E) is an interior IVIFS set of the complement of (f_A, E).

Definition 5.29 Let $((\xi_A, E), \tau)$ be an interval-valued intuitionistic fuzzy soft topological space on (ξ_A, E) and (f_A, E) be an IVIFS set in $P(\xi_A, E)$. Then, the union of all exterior IVIFS set of (f_A, E) is called the exterior of (f_A, E) and is denoted by $\text{ext}(f_A, E)$ and defined by $\text{ext}(f_A, E) = \cup\{(g_A, E)|(f_A, E)^c$ is a neighbourhood of $(g_A, E)\}$.

Clearly, from definition

$$\text{ext}(f_A, E) = \text{int}((f_A, E)^c).$$

Proposition 5.30 *For any two IVIFS sets (f_A, E) and (g_A, E) in an interval-valued intuitionistic fuzzy soft topological space $((\xi_A, E), \tau)$ on $P(\xi_A, E)$, then*

(i) $\text{ext}(f_A, E)$ *is open and is the largest open set contained in* $(f_A, E)^c$.
(ii) $(f_A, E)^c$ *is open if and only if* $(f_A, E)^c = \text{ext}(f_A, E)$.
(iii) $(g_A, E) \subseteq (f_A, E) \Rightarrow \text{ext}(f_A, E) \subseteq \text{ext}(g_A, E)$
(iv) $\text{ext}((g_A, E) \cap (f_A, E)) \supseteq \text{ext}(g_A, E) \cup \text{ext}(f_A, E)$
(v) $\text{ext}((g_A, E) \cup (f_A, E)) = \text{ext}(g_A, E) \cap \text{ext}(f_A, E)$

Proof Straight forward. $\square$

Definition 5.31 Let $((\xi_A, E), \tau)$ be an interval-valued intuitionistic fuzzy soft topological space on (ξ_A, E) and (f_A, E) be an IVIFS set in $P(\xi_A, E)$. Then, the intersection of all closed IVIFS set containing (f_A, E) is called the closure of (f_A, E) and is denoted by $\text{cl}(f_A, E)$ and defined by $\text{cl}(f_A, E) = \cap\{(g_A, E)|(g_A, E)$ is a IVIFS-closed set containing $(f_A, E)\}$.

Observe first that $\text{cl}(f_A, E)$ is an IVIFS closed set, since it is the intersection of IVIFS closed sets. Furthermore, $\text{cl}(f_A, E)$ is the smallest IVIFS closed set containing (f_A, E).

Example 5.32 Let us consider an interval-valued intuitionistic fuzzy soft topology $\tau_1 = \{(\phi_{\xi_A}, E), (\xi_A, E), (f_A^1, E), (f_A^2, E), (f_A^3, E), (f_A^4, E)\}$ as in the Example 5.6 and let

$$(f_A, E) = \left\{ e_1 = \{ u^1_{([0.2,0.3],[0.6,0.7])}, u^2_{([0,0.1],[0.4,0.5])}, u^3_{([0,0],[1,1])} \}, \right.$$

$$e_2 = \{ u^1_{([0.1,0.2],[0.5,0.6])}, u^2_{([0,0],[0.4,0.5])}, u^3_{([1,1],[0,0])} \},$$

$$e_3 = \{ u^1_{([0,0],[1,1])}, u^2_{([0,0],[1,1])}, u^3_{([0,0],[1,1])} \},$$

$$\left. e_4 = \{ u^1_{([0,0],[1,1])}, u^2_{([0,0],[1,1])}, u^3_{([0,0],[1,1])} \} \right\}$$

be an IVIFS set, and then,

$$\mathrm{cl}(f_A, E) = \cap \{ (g_A, E) | (g_A, E) \text{ is a IVIFS-closed set containing } (f_A, E) \}$$
$$= (f^1_A, E)^c \cap (f^3_A, E)^c$$
$$= (f^1_A, E)^c$$
$$= \left\{ e_1 = \{ u^1_{([0.2,0.3],[0.5,0.6])}, u^2_{([0,0.1],[0.4,0.5])}, u^3_{([0,0],[1,1])} \}, \right.$$

$$e_2 = \{ u^1_{([0.2,0.3],[0.4,0.5])}, u^2_{([0,0],[0.4,0.5])}, u^3_{([1,1],[0,0])} \},$$

$$\left. e_3 = \{ u^1_{([1,1],[0,0])}, u^2_{([1,1],[0,0])}, u^3_{([1,1],[0,0])} \} \right\}$$

Since $(f_A, E) \subseteq (f^1_A, E)^c$ and $(f_A, E) \subseteq (f^4_A, E)^c$.

Proposition 5.33 *For any two IVIFS sets (f_A, E) and (g_A, E) in an interval-valued intuitionistic fuzzy soft topological space $((\xi_A, E), \tau)$ on $P(\xi_A, E)$, then*

   (i) *$\mathrm{cl}(f_A, E)$ is the smallest IVIFS closed set containing (f_A, E).*
  (ii) *(f_A, E) is IVIFS closed if and only if $(f_A, E) = \mathrm{cl}(f_A, E)$.*
 (iii) *$(g_A, E) \subseteq (f_A, E) \Rightarrow \mathrm{cl}(g_A, E) \subseteq \mathrm{cl}(f_A, E)$*
  (iv) *$\mathrm{cl}(\mathrm{cl}(f_A, E)) = \mathrm{cl}(f_A, E)$*
   (v) *$\mathrm{cl}(\phi_{\xi_A}, E) = (\phi_{\xi_A}, E)$ and $\mathrm{cl}(\xi_A, E) = (\xi_A, E)$*
  (vi) *$\mathrm{cl}((g_A, E) \cup (f_A, E)) = \mathrm{cl}(g_A, E) \cup \mathrm{cl}(f_A, E)$*
 (vii) *$\mathrm{cl}((g_A, E) \cap (f_A, E)) \subseteq \mathrm{cl}(g_A, E) \cap \mathrm{cl}(f_A, E)$*

Proof Since $(g_A, E) \subseteq (f_A, E)$, implies all the closed set containing (f_A, E) also contained (g_A, E). Therefore,

   (i) $\cap \{ (g^*_A, E) | (g^*_A, E)$ is an IVIFS-closed set containing $(g_A, E) \} \subseteq \cap \{ (f^*_A, E) |$ (f^*_A, E) is an IVIFS-closed set containing $(f_A, E) \} \Rightarrow \mathrm{cl}(g_A, E) \subseteq \mathrm{cl}(f_A, E)$
  (ii) $\mathrm{cl}(\mathrm{cl}(f_A, E)) = \cup \{ (g_A, E) | (g_A, E)$ is an IVIFS-closed set containing $\mathrm{cl}(f_A, E) \}$, and since $\mathrm{cl}(f_A, E)$ is the largest closed IVIFS set containing $\mathrm{cl}(f_A, E)$, there- fore $\mathrm{cl}(\mathrm{cl}(f_A, E)) = \mathrm{cl}(f_A, E)$.
 (iii) Since $\mathrm{cl}((g_A, E)) \supseteq (g_A, E)$ and $\mathrm{cl}((f_A, E)) \supseteq (f_A, E)$

$$\Rightarrow \mathrm{cl}(g_A, E) \cup \mathrm{cl}(f_A, E) \supseteq (g_A, E) \cup (f_A, E)$$
$$\Rightarrow \mathrm{cl}(g_A, E) \cup \mathrm{cl}(f_A, E) \supseteq \mathrm{cl}((g_A, E) \cup (f_A, E)) \tag{5.3}$$

Again since $(g_A, E) \cup (f_A, E) \supseteq (g_A, E)$ and $(g_A, E) \cup (f_A, E) \supseteq (f_A, E)$

$$\text{cl}((g_A, E) \cup (f_A, E)) \supseteq \text{cl}(g_A, E) \text{ and cl}((g_A, E) \cup (f_A, E)) \supseteq \text{cl}(f_A, E)$$
$$\Rightarrow \text{cl}((g_A, E) \cup (f_A, E)) \supseteq \text{cl}(g_A, E) \cup \text{cl}(f_A, E) \tag{5.4}$$

Using (5.3) and (5.4), we get $\text{cl}((g_A, E) \cup (f_A, E)) = \text{cl}(g_A, E) \cup \text{cl}(f_A, E)$

(iv) Since

$$(g_A, E) \supseteq (g_A, E) \cap (f_A, E) \text{ and } (f_A, E) \supseteq (g_A, E) \cap (f_A, E)$$
$$\Rightarrow \text{cl}(g_A, E) \supseteq ((g_A, E) \cap (f_A, E)) \text{ and cl}(f_A, E) \supseteq \text{cl}((g_A, E) \cap (f_A, E))$$
$$\Rightarrow \text{cl}(g_A, E) \cap \text{cl}(f_A, E) \supseteq \text{cl}((g_A, E) \cap (f_A, E)).$$

$\square$

5.3 Interval-Valued Intuitionistic Fuzzy Soft Subspace Topology

Theorem 5.34 *Let* $((\xi_A, E), \tau)$ *be an interval-valued intuitionistic fuzzy soft topological space on* (ξ_A, E) *and* (f_A, E) *be an IVIFS set in* $P(\xi_A, E)$. *Then, the collection* $\tau_{(f_A, E)} = \{(f_A, E) \cap (g_A, E) | (g_A, E) \in \tau\}$ *is an interval-valued intuitionistic fuzzy soft topology on the interval-valued intuitionistic fuzzy soft set* (f_A, E).

Proof

(i) Since $(\phi_{\xi_A}, E), (\xi_A, E) \in \tau$, $(f_A, E) = (f_A, E) \cap (\xi_A, E)$, and $(\phi_{f_A}, E) = (f_A, E) \cap (\phi_{\xi_A}, E)$, therefore $(\phi_{f_A}, E), (f_A, E) \in \tau_{(f_A, E)}$.

(ii) Let $\{(f_A^i, E) | i = 1, 2, 3, \ldots, n\}$ be a finite subfamily of intuitionistic fuzzy soft open sets in $\tau_{(f_A, E)}$, and then for each $i = 1, 2, 3, \ldots, n$, there exist $(g_A^i, E) \in \tau$ such that $(f_A^i, E) = (f_A, E) \cap (g_A^i, E)$.
Now, $\cap_{i=1}^n (f_A^i, E) = \cap_{i=1}^n ((f_A, E) \cap (g_A^i, E)) = (f_A, E) \cap (\cap_{i=1}^n (g_A^i, E))$ and since $\cap_{i=1}^n (g_A^i, E) \in \tau \Rightarrow \cap_{i=1}^n (f_A^i, E) \in \tau_{(f_A, E)}$.

(iii) Let $\{(f_A^k, E) | k \in K\}$ be an arbitrary family of interval-valued intuitionistic fuzzy soft open sets in $\tau_{(f_A, E)}$, and then for each $k \in K$, there exist $(g_A^k, E) \in \tau$ such that $(f_A^k, E) = (f_A, E) \cap (g_A^k, E)$.
Now, $\cup_{k \in K} (f_A^k, E) = \cup_{k \in K} ((f_A, E) \cap (g_A^k, E)) = (f_A, E) \cap (\cup_{k \in K} (g_A^k, E))$ and since $\cup_{k \in K} (g_A^k, E) \in \tau \Rightarrow \cup_{k \in K} (f_A^k, E) \in \tau_{(f_A, E)}$. $\square$

Definition 5.35 Let $((\xi_A, E), \tau)$ be an IVIFS topological space on (ξ_A, E) and (f_A, E) be an IVIFS set in $P(\xi_A, E)$. Then, the IVIFS topology $\tau_{(f_A, E)} = \{(f_A, E) \cap (g_A, E) | (g_A, E) \in \tau\}$ is called interval-valued intuitionistic fuzzy soft subspace topology (in short and $((f_A, E), \tau_{(f_A, E)})$ is called interval-valued intuitionistic fuzzy soft subspace of $((\xi_A, E), \tau)$.

Example 5.36 Let us consider the interval-valued intuitionistic fuzzy soft topology $\tau_1 = \{(\phi_{\xi_A}, E), (\xi_A, E), (f_A^1, E), (f_A^2, E), (f_A^3, E), (f_A^4, E)\}$ as in the Example 5.6 and an IVIFS set

$$(f_A, E) = \Big\{ e_1 = \{ u^1_{([0.2,0.3],[0,0.1])}, u^2_{([0.5,0.6],[0.1,0.2])}, u^3_{([0.2,0.3],[0.6,0.7])} \},$$

$$e_2 = \{ u^1_{([0.3,0.4],[0.1,0.2])}, u^2_{([0.4,0.5],[0.2,0.3])}, u^3_{([0.4,0.5],[0.2,0.3])} \},$$

$$e_3 = \{ u^1_{([0,0],[1,1])}, u^2_{([0,0],[1,1])}, u^3_{([0,0],[1,1])} \} \Big\} \in P(\xi_A, E)$$

Then,

$$(\phi_{f_A}, E) = (f_A, E) \cap (\phi_{\xi_A}, E) = \Big\{ e_1 = \{ u^1_{([0,0],[1,1])}, u^2_{([0,0],[1,1])}, u^3_{([0,0],[1,1])} \},$$

$$e_2 = \{ u^1_{([0,0],[1,1])}, u^2_{([0,0],[1,1])}, u^3_{([0,0],[1,1])} \},$$

$$e_3 = \{ u^1_{([0,0],[1,1])}, u^2_{([0,0],[1,1])}, u^3_{([0,0],[1,1])} \} \Big\}$$

$$(g_A^1, E) = (f_A, E) \cap (f_A^1, E) = \Big\{ e_1 = \{ u^1_{([0.2,0.3],[0.2,0.3])}, u^2_{([0.4,0.5],[0.1,0.2])},$$

$$u^3_{([0.2,0.3],[0.6,0.7])} \},$$

$$e_2 = \{ u^1_{([0.3,0.4],[0.2,0.3])}, u^2_{([0.4,0.5],[0.2,0.3])}, u^3_{([0,0],[1,1])} \},$$

$$e_3 = \{ u^1_{([0,0],[1,1])}, u^2_{([0,0],[1,1])}, u^3_{([0,0],[1,1])} \} \Big\}$$

$$(g_A^2, E) = (f_A, E) \cap (f_A^2, E) = \Big\{ e_1 = \{ u^1_{([0.2,0.3],[0.1,0.2])}, u^2_{([0.5,0.6],[0.2,0.3])},$$

$$u^3_{([0.2,0.3],[0.6,0.7])} \},$$

$$e_2 = \{ u^1_{([0.2,0.3],[0.1,0.2])}, u^2_{([0.4,0.5],[0.2,0.3])}, u^3_{([0,0],[1,1])} \},$$

$$e_3 = \{ u^1_{([0,0],[1,1])}, u^2_{([0,0],[1,1])}, u^3_{([0,0],[1,1])} \} \Big\}$$

$$(g_A^3, E) = (f_A, E) \cap (f_A^3, E) - \Big\{ e_1 = \{ u^1_{([0.2,0.3],[0.2,0.3])}, u^2_{([0.4,0.5],[0.2,0.3])},$$

$$u^3_{([0.2,0.3],[0.6,0.7])} \},$$

$$e_2 = \{ u^1_{([0.2,0.3],[0.2,0.3])}, u^2_{([0.4,0.5],[0.2,0.3])}, u^3_{([0,0],[1,1])} \},$$

$$e_3 = \{ u^1_{([0,0],[1,1])}, u^2_{([0,0],[1,1])}, u^3_{([0,0],[1,1])} \} \Big\}$$

$$(g_A^4, E) = (f_A, E) \cap (f_A^4, E) = \Big\{ e_1 = \{ u^1_{([0.2,0.3],[0.1,0.2])}, u^2_{([0.5,0.6],[0.1,0.2])},$$

$$u^3_{([0.2,0.3],[0.6,0.7])} \},$$

$$e_2 = \{ u^1_{([0.3,0.4],[0.1,0.2])}, u^2_{([0.4,0.5],[0.2,0.3])}, u^3_{([0,0],[1,1])} \},$$

$$e_3 = \{ u^1_{([0,0],[1,1])}, u^2_{([0,0],[1,1])}, u^3_{([0,0],[1,1])} \} \Big\}$$

Then, $\tau_{(f_A,E)} = \left\{ \left(\phi_{f_A}, E \right), (f_A, E), \left(g_A^1, E \right), \left(g_A^2, E \right), \left(g_A^3, E \right), \left(g_A^4, E \right) \right\}$ is an interval-valued intuitionistic fuzzy soft subspace topology for τ and $\left((f_A, E), \tau_{(f_A,E)} \right)$ is called interval-valued intuitionistic fuzzy soft subspace of $((\xi_A, E), \tau_1)$.

Theorem 5.37 *Let* $((\xi_A, E), \tau)$ *be an interval-valued intuitionistic fuzzy soft topological space on* (ξ_A, E), *B be an interval-valued intuitionistic fuzzy soft basis for* τ, *and* (f_A, E) *be an IVIFS set in* $P(\xi_A, E)$. *Then, the family* $B_{(f_A,E)} = \{(f_A, E) \cap (g_A, E) | (g_A, E) \in B\}$ *is an interval-valued intuitionistic fuzzy soft basis for subspace topology* $\tau_{(f_A,E)}$.

Proof Let $(h_A, E) \in \tau_{(\xi_A,E)}$, and then, there exists an IVIFS set $(g_A, E) \in \tau$, such that $(h_A, E) = (f_A, E) \cap (g_A, E)$. Since B is a base for τ, there exists subcollection $\{(\chi_A^i, E) | i \in I\}$ of B such that $(g_A, E) = \cup_{i \in I}(\chi_A^i, E)$. Therefore, $(h_A, E) = (f_A, E) \cap (g_A, E) = (f_A, E) \cap (\cup_{i \in I}(\chi_A^i, E)) = \cup_{i \in I}((f_A, E) \cap (\chi_A^i, E))$. Since $(f_A, E) \cap \left(\chi_A^i, E \right) \in B_{(f_A,E)}$, which implies that $B_{(f_A,E)}$ is an IVIFS basis for the IVIFS subspace topology $\tau_{(f_A,E)}$. $\square$

Theorem 5.38 *Let* $((\xi_A, E), \tau)$ *be an IVIFS topological subspace of* $((\eta_A, E), \tau^*)$ *and let* $((\eta_A, E), \tau^*)$ *be an IVIFS topological subspace of* $((\Im_A, E), \tau^{**})$. *Then,* $((\xi_A, E), \tau)$ *is also an IVIFS topological subspace of* $((\Im_A, E), \tau^{**})$.

Proof Since $(\xi_A, E) \subseteq (\eta_A, E) \subseteq (\Im_A, E)$, $((\xi_A, E), \tau)$ is an interval-valued intuitionistic fuzzy soft topological subspace of $((\Im_A, E), \tau^{**})$ if and only if $\tau_{(\xi_A,E)}^{**} = \tau$. Let $(f_A, E) \in \tau$; now, since $((\xi_A, E), \tau)$ be an interval-valued intuitionistic fuzzy soft topological subspace of $((\eta_A, E), \tau^*)$, i.e. $\tau_{(\eta_A,E)}^* = \tau$, there exist $(f_A^*, E) \in \tau^*$ such that $(f_A, E) = (\xi_A, E) \cap (f_A^*, E)$. But $((\eta_A, E), \tau^*)$ be an interval-valued intuitionistic fuzzy soft topological subspace of $((\Im_A, E), \tau^{**})$, there exist $(f_A^{**}, E) \in \tau^{**}$ such that $(f_A^*, E) = (\eta_A, E) \cap (f_A^{**}, E)$. Thus, $(f_A, E) = (\xi_A, E) \cap (f_A^*, E) = (\xi_A, E) \cap (\eta_A, E) \cap (f_A^{**}, E) = (\xi_A, E) \cap (f_A^{**}, E)$, and since $(\xi_A, E) \subseteq (\eta_A, E)$, $(f_A, E) \in \tau_{(\xi_A,E)}^{**}$. Accordingly, $\tau \subseteq \tau_{(\xi_A,E)}^{**}$.

Now, assume $(g_A, E) \in \tau_{(\xi_A,E)}^{**}$, i.e., there exist $(h_A, E) \in \tau^{**}$ such that $(g_A, E) = (\xi_A, E) \cap (h_A, E)$. But $(\eta_A, E) \cap (h_A, E) \in \tau_{(\eta_A,E)}^{**} = \tau^*$, $(\xi_A, E) \cap ((\eta_A, E) \cap (h_A, E)) \in \tau_{(\xi_A,E)}^* = \tau$. Since $(\xi_A, E) \cap ((\eta_A, E) \cap (h_A, E)) = (\xi_A, E) \cap (h_A, E) = (g_A, E)$, we have $(g_A, E) \in \tau$. Accordingly, $\tau_{(\xi_A,E)}^{**} \subseteq \tau$, and thus, the theorem is proved. $\square$

5.4 Conclusion

Soft sets, fuzzy soft sets, intuitionistic fuzzy soft sets, and interval-valued intuitionistic fuzzy soft sets are all mathematical tools for dealing with uncertainties. In this chapter, we have introduced the concept of interval-valued intuitionistic fuzzy soft topological spaces together with some basic concepts over a fixed parameter set, which is the extension of fuzzy soft topological spaces introduced by Tugbahan et al.

References

1. Ali, M.I., Feng, F., Liu, X., Min, W.K., Shabir, M.: On some new operations in soft set theory. Comput. Math Appl. **57**(9), 1547–1553 (2009)
2. Atanassov, K.: Intuitionistic fuzzy sets. Fuzzy Sets Syst. **20**, 87–96 (1986)
3. Atanassov, K., Gargov, G.: Interval-valued intuitionistic fuzzy sets. Fuzzy Sets Syst. **31**, 343–349 (1989)
4. Jiang, Y., Tang, Y., Chen, Q., Liu, H., Tung, J.: Interval-valued intuitionistic fuzzy soft sets and their properties. Comput. Math Appl. **60**, 906–918 (2010)
5. Maji, P.K., Biswas, R., Roy, A.R.: Soft set theory. Comput. Math Appl. **45**(555–562), 191–209 (2003)
6. Maji, P.K., Roy, A.R., Biswas, R.: Fuzzy soft sets. J. Fuzzy Math. **9**(3), 589–602 (2001)
7. Maji, P.K., Biswas, R., Roy, A.R.: Intuitionistic fuzzy soft sets. J. Fuzzy Math. **12**(3), 677–692 (2004)
8. Molodtsov, D.: Soft set theory-first results. Comput. Math Appl. **37**(4–5), 19–31 (1999)
9. Roy, S., Samanta, T.K.: A note on fuzzy soft topological. Ann. Fuzzy Math. Inform. (2011)
10. Shabir, M., Naz, M.: On soft topological spaces. Comput. Math Appl. **61**, 1786–1799 (2011)
11. Simsekler, T., Yuksel, S.: Fuzzy soft topological spaces. Ann. Fuzzy Math. Inform. **3**, 305–311 (2011)
12. Tanay, B., Kandemir, M.B.: Topological structure of fuzzy soft sets. Comput. Math Appl. **61**, 2952–2957 (2011)
13. Zadeh, L.A.: Fuzzy sets. Inform. Control **8**, 338–353 (1965)
14. Zhaowen, L., Rongchen, C.: On the topological structure of intuitionistic fuzzy soft sets. Ann. Fuzzy Math. Inform. **5**, 229–239 (2013)

References

Chapter 6
Interval-Valued Intuitionistic Fuzzy Soft Multi-Sets and Their Relations

Abstract In this chapter, we introduce the concept of interval-valued intuitionistic fuzzy soft multi-sets and study its properties and operations. Then, the concept of interval-valued intuitionistic fuzzy soft multi-set relations (IVIFSMS-relations) is proposed. The basic properties of the IVIFSMS-relations are also discussed. Finally, various types of IVIFSMS-relations are presented.

Keywords Fuzzy set · Soft set · Soft multi-set · Fuzzy soft multi-set · Interval-valued intuitionistic fuzzy soft multi-set · Interval-valued intuitionistic fuzzy soft multi-set relation

Most of the problems in engineering, medical science, economics, environments, etc. have various uncertainties. Molodtsov [9] initiated the concept of soft set theory as a mathematical tool for dealing with uncertainties. Research works on soft set theory are progressing rapidly. Maji et al. [6] defined several operations on soft set theory. Based on the analysis of several operations on soft sets introduced in [6], Ali et al. [1] presented some new algebraic operations for soft sets. Combining soft sets with fuzzy sets [10] and intuitionistic fuzzy sets [4], Maji et al. [7, 8] defined fuzzy soft sets and intuitionistic fuzzy soft sets which have rich potentials for solving decision-making problems. The notion of the interval-valued intuitionistic fuzzy set was introduced by Atanassov and Gargov [5]. Alkhazaleh et al. [2] as a generalisation of Molodtsov's soft set presented the definition of a soft multi-set and its basic operations such as complement, union, and intersection. In 2012, Alkhazaleh and Salleh [3] introduced the concept of fuzzy soft multi-set theory and studied the application of these sets.

In this chapter, we introduce the concept of interval-valued intuitionistic fuzzy soft multi-sets and study its properties and operations. Also the concept of interval-valued intuitionistic fuzzy soft multi-set relations (IVIFSMS-relations) is proposed. The basic properties of the IVIFSMS-relations are discussed. Also various types of IVIFSMS-relations are presented.

In this section, we introduce the definition of an interval-valued intuitionistic fuzzy soft multi-set, and its basic operations such as complement, union, and intersection. We give examples for these concepts. Basic properties of the operations are also given.

Definition 6.1 Let $\{U_i : i \in I\}$ be a collection of universes such that $\bigcap_{i \in I} U_i = \phi$ and let $\{E_{U_i} : i \in I\}$ be a collection of sets of parameters. Let $U = \prod_{i \in I} \text{IVIFS}(U_i)$ where $\text{IVIFS}(U_i)$ denotes the set of all interval-valued intuitionistic fuzzy subsets of U_i, $E = \prod_{i \in I} E_{U_i}$ and $A \subseteq E$. Then, the pair (F, A) is called an *interval-valued intuitionistic fuzzy soft multi-set* over U where F is a mapping given by $F : A \to U$.

Remark 6.2 It is to be noted that if $U = \prod_{i \in I} \text{IFS}(U_i)$, then the pair (F, A) is called an *intuitionistic fuzzy soft multi-set* over U. To illustrate this, let us consider the following example:

Let us consider there are three universes U_1, U_2, and U_3.

Let $U_1 = \{h_1, h_2, h_3\}$, $U_2 = \{c_1, c_2, c_3\}$, and $U_3 = \{v_1, v_2\}$. Let $\{E_{U_1}, E_{U_2}, E_{U_3}\}$ be a collection of sets of decision parameters related to the above universes where

$$E_{U_1} = \{e_{U_1,1}, e_{U_1,2}, e_{U_1,3}\}, \quad E_{U_2} = \{e_{U_2,1}, e_{U_2,2}, e_{U_2,3}\},$$
$$E_{U_3} = \{e_{U_3,1}, e_{U_3,2}, e_{U_3,3}\}.$$

Let $U = \prod_{i=1}^{3} \text{IFS}(U_i)$, $E = \prod_{i=1}^{3} E_{U_i}$, and $A \subseteq E$, such that

$$A = \{a_1 = (e_{U_1,1}, e_{U_2,1}, e_{U_3,1}), a_2 = (e_{U_1,1}, e_{U_2,2}, e_{U_3,1})\}$$

Let $F : A \to U$ be defined by

$$F(a_1) = (\{\langle h_1, 0.3, 0.5\rangle, \langle h_2, 0.4, 0.6\rangle, \langle h_3, 0.9, 0.1\rangle\},$$
$$\{\langle c_1, 0.9, 0\rangle, \langle c_2, 0.6, 0.3\rangle, \langle c_3, 0.5, 0.4\rangle\},$$
$$\{\langle v_1, 0.9, 0\rangle, \langle v_2, 0.8, 0.2\rangle\}),$$
$$F(a_2) = (\{\langle h_1, 0.4, 0.4\rangle, \langle h_2, 0.7, 0.3\rangle, \langle h_3, 0.6, 0.4\rangle\},$$
$$\{\langle c_1, 0.6, 0.3\rangle, \langle c_2, 0.4, 0.4\rangle, \langle c_3, 0.7, 0.2\rangle\},$$
$$\{\langle v_1, 0.6, 0.3\rangle, \langle v_2, 0.5, 0.3\rangle\}).$$

Then

$$(F, A) = \{(a_1, (\{\langle h_1, 0.3, 0.5\rangle, \langle h_2, 0.4, 0.6\rangle, \langle h_3, 0.9, 0.1\rangle\},$$
$$\{\langle c_1, 0.9, 0\rangle, \langle c_2, 0.6, 0.3\rangle, \langle c_3, 0.5, 0.4\rangle\},$$
$$\{\langle v_1, 0.9, 0\rangle, \langle v_2, 0.8, 0.2\rangle\})),$$
$$(a_2, (\{\langle h_1, 0.4, 0.4\rangle, \langle h_2, 0.7, 0.3\rangle, \langle h_3, 0.6, 0.4\rangle\},$$
$$\{\langle c_1, 0.6, 0.3\rangle, \langle c_2, 0.4, 0.4\rangle, \langle c_3, 0.7, 0.2\rangle\},$$
$$\{\langle v_1, 0.6, 0.3\rangle, \langle v_2, 0.5, 0.3\rangle\}))\}.$$

Here, (F, A) is an intuitionistic fuzzy soft multi-set over U.

Example 6.3 Let us consider there are three universes U_1, U_2, and U_3.

Let $U_1 = \{h_1, h_2, h_3\}$, $U_2 = \{c_1, c_2, c_3\}$, and $U_3 = \{v_1, v_2\}$. Let $\{E_{U_1}, E_{U_2}, E_{U_3}\}$ be a collection of sets of decision parameters related to the above universes where

$$E_{U_1} = \{e_{U_1,1}, e_{U_1,2}, e_{U_1,3}\}, \quad E_{U_2} = \{e_{U_2,1}, e_{U_2,2}, e_{U_2,3}\},$$
$$E_{U_3} = \{e_{U_3,1}, e_{U_3,2}, e_{U_3,3}\}.$$

Let $U = \prod_{i=1}^{3} \text{IVIFS}(U_i)$, $E = \prod_{i=1}^{3} E_{U_i}$, and $A \subseteq E$, such that

$$A = \{a_1 = (e_{U_1,1}, e_{U_2,1}, e_{U_3,1}), a_2 = (e_{U_1,1}, e_{U_2,2}, e_{U_3,1})\}$$

Let $F\!:\!A \to U$ be defined by

$$
\begin{aligned}
F(a_1) = (&\{\langle h_1, [0.1, 0.3], [0.3, 0.5]\rangle, \langle h_2, [0.3, 0.4], [0.4, 0.6]\rangle, \\
&\langle h_3, [0.7, 0.9], [0, 0.1]\rangle\}, \{\langle c_1, [0.7, 0.9], [0, 0]\rangle, \\
&\langle c_2, [0.4, 0.6], [0.2, 0.3]\rangle, \langle c_3, [0.3, 0.5], [0.2, 0.4]\rangle\}, \\
&\{\langle v_1, [0.7, 0.9], [0, 0]\rangle, \langle v_2, [0.6, 0.8], [0, 0.2]\rangle\}), \\
F(a_2) = (&\{\langle h_1, [0.2, 0.4], [0.3, 0.4]\rangle, \langle h_2, [0.6, 0.7], [0.2, 0.3]\rangle, \\
&\langle h_3, [0.5, 0.6], [0.3, 0.4]\rangle\}, \{\langle c_1, [0.3, 0.6], [0.2, 0.3]\rangle, \\
&\langle c_2, [0.1, 0.4], [0.3, 0.4]\rangle, \langle c_3, [0.5, 0.7], [0.1, 0.2]\rangle\}, \\
&\{\langle v_1, [0.5, 0.6], [0.2, 0.3]\rangle, \langle v_2, [0.2, 0.5], [0.1, 0.3]\rangle\})
\end{aligned}
$$

Then,

$$
\begin{aligned}
(F, A) = \{(&a_1, (\{\langle h_1, [0.1, 0.3], [0.3, 0.5]\rangle, \langle h_2, [0.3, 0.4], [0.4, 0.6]\rangle, \\
&\langle h_3, [0.7, 0.9], [0, 0.1]\rangle\}, \{\langle c_1, [0.7, 0.9], [0, 0]\rangle, \\
&\langle c_2, [0.4, 0.6], [0.2, 0.3]\rangle, \langle c_3, [0.3, 0.5], [0.2, 0.4]\rangle\}, \\
&\{\langle v_1, [0.7, 0.9], [0, 0]\rangle, \langle v_2, [0.6, 0.8], [0, 0.2]\rangle\})), \\
(&a_2, (\{\langle h_1, [0.2, 0.4], [0.3, 0.4]\rangle, \langle h_2, [0.6, 0.7], [0.2, 0.3]\rangle, \\
&\langle h_3, [0.5, 0.6], [0.3, 0.4]\rangle\}, \{\langle c_1, [0.3, 0.6], [0.2, 0.3]\rangle, \\
&\langle c_2, [0.1, 0.4], [0.3, 0.4]\rangle, \langle c_3, [0.5, 0.7], [0.1, 0.2]\rangle\}, \\
&\{\langle v_1, [0.5, 0.6], [0.2, 0.3]\rangle, \langle v_2, [0.2, 0.5], [0.1, 0.3]\rangle\}))\}.
\end{aligned}
$$

Here, (F, A) is an interval-valued intuitionistic fuzzy soft multi-set over U.

Definition 6.4 Let (F, A) and (G, B) be two interval-valued intuitionistic fuzzy soft multi-sets over U. Then, (F, A) is called an *interval-valued intuitionistic fuzzy soft multi-subset* of (G, B) if

(a) $A \subseteq B$ and

(b) $\forall e_{U_{i,j}} \in a_k, \left(e_{U_{i,j}}, F_{e_{U_{i,j}}} \right)$ is an interval-valued intuitionistic fuzzy subset of $\left(e_{U_{i,j}}, G_{e_{U_{i,j}}} \right)$,

where $a_k \in A, k \in \{ 1,2,3,\ldots,n \}, i \in \{1,2,3,\ldots,m\}$ and $j \in \{1,2,3,\ldots,r\}$. This relationship is denoted by $(F,A)\tilde{\subseteq}(G,B)$.

Example 6.5 Let us consider Example 6.3.

Let $B = \{ b_1 = (e_{U_1,1}, e_{U_2,1}, e_{U_3,1}), b_2 = (e_{U_1,1}, e_{U_2,2}, e_{U_3,1}), b_3 = (e_{U_1,2}, e_{U_2,2}, e_{U_3,2}) \}$.

Clearly $A \subseteq B$. Let (G, B) be two interval-valued intuitionistic fuzzy soft multi-set over U, such that

$$
\begin{aligned}
(G,B) = \{ &(b_1,(\{\langle h_1, [0.2,0.4], [0.3,0.5]\rangle, \langle h_2, [0.4,0.6], [0.2,0.4]\rangle, \\
&\langle h_3, [0.7,0.9], [0,0.1]\rangle\}, \{\langle c_1, [0.7,0.9], [0,0]\rangle, \\
&\langle c_2, [0.6,0.8], [0.1,0.2]\rangle, \langle c_3, [0.4,0.6], [0.2,0.4]\rangle\}, \\
&\{\langle v_1, [0.7,0.9], [0,0]\rangle, \langle v_2, [0.7,0.9], [0,0.1]\rangle\})), \\
&(b_2, (\{\langle h_1, [0.3,0.5], [0.2,0.3]\rangle, \langle h_2, [0.6,0.8], \\
&[0.1,0.2]\rangle, \langle h_3, [0.7,0.9], [0,0.1]\rangle\}\{\langle c_1, [0.4,0.7], [0.1,0.2]\rangle, \\
&\langle c_2, [0.4,0.6], [0.2,0.3]\rangle, \langle c_3, [0.7,0.9], [0,0]\rangle\}, \\
&\{\langle v_1, [0.5,0.7], [0.2,0.3]\rangle, \langle v_2, [0.3,0.5], [0.1,0.3]\rangle\})), \\
&(b_3, (\{\langle h_1, [0.6,0.8], [0.1,0.2]\rangle, \langle h_2, [0.6,0.8], [0.1,0.2]\rangle, \\
&\langle h_3, [0,0.2], [0.5,0.7]\rangle\{\langle c_1, [0.7,0.9], [0,0.1]\rangle, \\
&\langle c_2, [0.5,0.7], [0.1,0.3]\rangle, \langle c_3, [0.2,0.4], [0.3,0.5]\rangle\}, \\
&\{\langle v_1, [0.4,0.6], [0.2,0.4]\rangle, \langle v_2, [0.3,0.5], [0.2,0.4]\rangle\})),
\end{aligned}
$$

Therefore $(F,A) \tilde{\subseteq} (G,B)$.

Definition 6.6 The *complement* of an interval-valued intuitionistic fuzzy soft multi-set (F, A) over U is denoted by $(F,A)^c$ and is defined by $(F,A)^c = (F^c,A)$, where $F^c:A \rightarrow U$ is a mapping given by $F^c(\alpha) = c(F(\alpha)), \forall \alpha \in A$ where 'c' is the interval-valued intuitionistic fuzzy complement.

Example 6.7 Let us consider Example 6.3. Then,

$$
\begin{aligned}
(F,A)^c = \{ &(a_1,(\{\langle h_1, [0.3,0.5], [0.1,0.3]\rangle, \langle h_2, [0.4,0.6], [0.3,0.4]\rangle, \\
&\langle h_3, [0.0,0.1], [0.7,0.9]\rangle, \{\langle c_1, [0.0,0.0], [0.7,0.9]\rangle, \\
&\{\langle c_2, [0.2,0.3], [0.4,0.6]\rangle, \langle c_3, [0.2,0.4], [0.3,0.5]\rangle\}, \\
&\{\langle v_1, [0.0,0.0], [0.7,0.9]\rangle, \langle v_2, [0.0,0.2], [0.6,0.8]\rangle\})), \\
&(a_2,(\{\langle h_1, [0.3,0.4], [0.2,0.4]\rangle, \langle h_2, [0.2,0.3], [0.6,0.7]\rangle, \langle h_3, [0.3,0.4], [0.5,0.6]\rangle\}, \\
&\{\langle c_1, [0.2,0.3], [0.3,0.6]\rangle, \langle c_2, [0.3,0.4], [0.1,0.4]\rangle, \langle c_3, [0.1,0.2], [0.5,0.7]\rangle\}, \\
&\{\langle v_1, [0.2,0.3], [0.5,0.6]\rangle, \langle v_2, [0.1,0.3], [0.2,0.5]\rangle\}))\}.
\end{aligned}
$$

Definition 6.8 An interval-valued intuitionistic fuzzy soft multi-set (F, A) over U is called a *semi-null interval-valued intuitionistic fuzzy soft multi-set*, denoted by $(F, A)_{\approx \phi}$, if at least one of an interval-valued intuitionistic fuzzy soft multi-set parts of (F, A) equals ϕ.

Example 6.9 Let us consider there are three universes U_1, U_2, and U_3. Let $U_1 = \{h_1, h_2, h_3\}$, $U_2 = \{c_1, c_2, c_3\}$, and $U_3 = \{v_1, v_2\}$. Let $\{E_{U_1}, E_{U_2}, E_{U_3}\}$ be a collection of sets of decision parameters related to the above universes where

$$E_{U_1} = \{e_{U_1,1}, e_{U_1,2}, e_{U_1,3}\}, \quad E_{U_2} = \{e_{U_2,1}, e_{U_2,2}, e_{U_2,3}\},$$
$$E_{U_3} = \{e_{U_3,1}, e_{U_3,2}, e_{U_3,3}\}.$$

Let $U = \prod_{i=1}^{3} \text{IVIFS}(U_i)$, $E = \prod_{i=1}^{3} E_{U_i}$, and $A \subseteq E$, such that

$$A = \{a_1 = (e_{U_1,1}, e_{U_2,1}, e_{U_3,1}), a_2 = (e_{U_1,1}, e_{U_2,2}, e_{U_3,1})\}$$

Then, a semi-null interval-valued intuitionistic fuzzy soft multi-set $(F, A)_{\approx \phi}$ is given by

$$(F, A)_{\approx \varphi} = \{(a_1, (\{\langle h_1, [0, 0], [1, 1]\rangle, \langle h_2, [0, 0], [1, 1]\rangle, \langle h_3, [0, 0], [1, 1]\rangle\},$$
$$\{\langle c_1, [0.7, 0.9], [0, 0]\rangle, \langle c_2, [0.4, 0.6], [0.3, 0.4]\rangle,$$
$$\langle c_3, [0.3, 0.5], [0.2, 0.4]\rangle\}, \{\langle v_1, [0.7, 0.9], [0, 0]\rangle,$$
$$\langle v_2, [0.6, 0.8], [0, 0.2]\rangle\})), (a_2, (\{\langle h_1, [0, 0], [1, 1]\rangle,$$
$$\langle h_2, [0, 0], [1, 1]\rangle, \langle h_3, [0, 0], [1, 1]\rangle\}, \{\langle c_1, [0.3, 0.5], [0.2, 0.4]\rangle,$$
$$\langle c_2, [0.4, 0.5], [0.3, 0.5]\rangle, \langle c_3, [0.7, 0.9], [0, 0]\rangle\},$$
$$\{\langle v_1, [0.3, 0.5], [0.2, 0.4]\rangle, \langle v_2, [0.3, 0.5], [0.2, 0.4]\rangle\})))\}.$$

Definition 6.10 An interval-valued intuitionistic fuzzy soft multi-set (F, A) over U is called a *null interval-valued intuitionistic fuzzy soft multi-set*, denoted by $(F, A)_{\phi}$, if all the interval-valued intuitionistic fuzzy soft multi-set parts of (F, A) equals ϕ.

Example 6.11 Let us consider there are three universes U_1, U_2, and U_3.
Let $U_1 = \{h_1, h_2, h_3\}$, $U_2 = \{c_1, c_2, c_3\}$, and $U_3 = \{v_1, v_2\}$. Let $\{E_{U_1}, E_{U_2}, E_{U_3}\}$ be a collection of sets of decision parameters related to the above universes where

$$E_{U_1} = \{e_{U_1,1}, e_{U_1,2}, e_{U_1,3}\}, \quad E_{U_2} = \{e_{U_2,1}, e_{U_2,2}, e_{U_2,3}\},$$
$$E_{U_3} = \{e_{U_3,1}, e_{U_3,2}, e_{U_3,3}\}.$$

Let $U = \prod_{i=1}^{3} \text{IVIFS}(U_i)$, $E = \prod_{i=1}^{3} E_{U_i}$, and $A \subseteq E$, such that

$$A = \{a_1 = (e_{U_1,1}, e_{U_2,1}, e_{U_3,1}), a_2 = (e_{U_1,1}, e_{U_2,2}, e_{U_3,1})\}.$$

Then, a null interval-valued intuitionistic fuzzy soft multi-set $(F, A)_\phi$ is given by

$$(F, A)_\varphi = \{(a_1, (\{\langle h_1, [0, 0], [1, 1]\rangle, \langle h_2, [0, 0], [1, 1]\rangle, \langle h_3, [0, 0], [1, 1]\rangle\},$$
$$\{\langle c_1, [0, 0], [1, 1]\rangle, \langle c_2, [0, 0], [1, 1]\rangle, \langle c_3, [0, 0], [1, 1]\rangle\},$$
$$\{\langle v_1, [0, 0], [1, 1]\rangle, \langle v_2, [0, 0], [1, 1]\rangle\})),$$
$$\{\langle c_1, [0, 0], [1, 1]\rangle, \langle c_2, [0, 0], [1, 1]\rangle, \langle c_3, [0, 0], [1, 1]\rangle\},$$
$$\{\langle v_1, [0, 0], [1, 1]\rangle, \langle v_2, [0, 0], [1, 1]\rangle\}))\}.$$

Definition 6.12 An interval-valued intuitionistic fuzzy soft multi-set (F, A) over U is called a *semi-absolute interval-valued intuitionistic fuzzy soft multi-set*, denoted by $(F, A)_{\approx U_i}$, if $(e_{U_i,j}, F_{e_{U_i,j}}) = U_i$ for at least one i, $a_k \in A$, $k = \{1, 2, 3, \ldots, n\}$, $i \in \{1, 2, 3, \ldots, m\}$, and $j \in \{1, 2, 3, \ldots, r\}$.

Example 6.13 Let us consider there are three universes U_1, U_2, and U_3.
Let $U_1 = \{h_1, h_2, h_3\}$, $U_2 = \{c_1, c_2, c_3\}$, and $U_3 = \{v_1, v_2\}$. Let $\{E_{U_1}, E_{U_2}, E_{U_3}\}$ be a collection of sets of decision parameters related to the above universes where

$$E_{U_1} = \{e_{U_1,1}, e_{U_1,2}, e_{U_1,3}\}, \quad E_{U_2} = \{e_{U_2,1}, e_{U_2,2}, e_{U_2,3}\},$$
$$E_{U_3} = \{e_{U_3,1}, e_{U_3,2}, e_{U_3,3}\}.$$

Let $U = \prod_{i=1}^{3} \text{IVIFS}(U_i)$, $E = \prod_{i=1}^{3} E_{U_i}$, and $A \subseteq E$, such that $A = \{a_1 = (e_{U_1,1}, e_{U_2,1}, e_{U_3,1}), a_2 = (e_{U_1,1}, e_{U_2,2}, e_{U_3,1})\}$. Then, the semi-absolute interval-valued intuitionistic fuzzy soft multi-set $(F, A)_{\approx U_i}$ is given by

$$(F, A)_{\approx U_i} = \{(a_1, (\{\langle h_1, [1, 1], [0, 0]\rangle, \langle h_2, [1, 1], [0, 0]\rangle, \langle h_3, [1, 1], [0, 0]\rangle\},$$
$$\{\langle c_1, [0.7, 0.9], [0, 0]\rangle, \langle c_2, [0.4, 0.5], [0.3, 0.5]\rangle,$$
$$\langle c_3, [0.3, 0.5], [0.2, 0.4]\rangle\}, \{\langle v_1, [0.7, 0.9], [0, 0]\rangle,$$
$$\langle v_2, [0.6, 0.8], [0, 0.2]\rangle\})), (a_2, (\{\langle h_1, [1, 1], [0, 0]\rangle,$$
$$\langle h_2, [1, 1], [0, 0]\rangle, \langle h_3, [1, 1], [0, 0]\rangle\}, \{\langle c_1, [0.3, 0.5], [0.2, 0.4]\rangle,$$
$$\langle c_2, [0.4, 0.6], [0.3, 0.4]\rangle, \langle c_3, [0.7, 0.9], [0, 0]\rangle\},$$
$$\{\langle v_1, [0.3, 0.5], [0.2, 0.4]\rangle, \langle v_2, [0.3, 0.5], [0.2, 0.4]\rangle\}))\}.$$

Definition 6.14 An interval-valued intuitionistic fuzzy soft multi-set (F, A) over U is called an *absolute interval-valued intuitionistic fuzzy soft multi-set*, denoted by $(F, A)_U$, if $(e_{U_i,j}, F_{e_{U_i,j}}) = U_i$, $\forall i$.

Example 6.15 Let us consider there are three universes U_1, U_2, and U_3.
Let $U_1 = \{h_1, h_2, h_3\}$, $U_2 = \{c_1, c_2, c_3\}$, and $U_3 = \{v_1, v_2\}$. Let $\{E_{U_1}, E_{U_2}, E_{U_3}\}$ be a collection of sets of decision parameters related to the above universes where

$$E_{U_1} = \{e_{U_1,1}, e_{U_1,2}, e_{U_1,3}\}, \quad E_{U_2} = \{e_{U_2,1}, e_{U_2,2}, e_{U_2,3}\},$$
$$E_{U_3} = \{e_{U_3,1}, e_{U_3,2}, e_{U_3,3}\}.$$

Let $U = \prod_{i=1}^{3} \text{IVIFS}(U_i)$, $E = \prod_{i=1}^{3} E_{U_i}$, and $A \subseteq E$, such that $A = \{a_1 = (e_{U_1,1}, e_{U_2,1}, e_{U_3,1}), a_2 = (e_{U_1,1}, e_{U_2,2}, e_{U_3,1})\}$. Then, an absolute interval-valued intuitionistic fuzzy soft multi-set $(F, A)_U$ is given by

$$(F, A)_U = \{(a_1, (\{\langle h_1, [1, 1], [0, 0]\rangle, \langle h_2, [1, 1], [0, 0]\rangle, \langle h_3, [1, 1], [0, 0]\rangle\},$$
$$\{\langle c_1, [1, 1], [0, 0]\rangle, \langle c_2, [1, 1], [0, 0]\rangle, \langle c_3, [1, 1], [0, 0]\rangle\},$$
$$\{\langle v_1, [1, 1], [0, 0]\rangle, \langle v_2, [1, 1], [0, 0]\rangle\})),$$
$$\{(a_2, (\{\langle h_1, [1, 1], [0, 0]\rangle, \langle h_2, [1, 1], [0, 0]\rangle, \langle h_3, [1, 1], [0, 0]\rangle\},$$
$$\{\langle c_1, [1, 1], [0, 0]\rangle, \langle c_2, [1, 1], [0, 0]\rangle, \langle c_3, [1, 1], [0, 0]\rangle\},$$
$$\{\langle v_1, [1, 1], [0, 0]\rangle, \langle v_2, [1, 1], [0, 0]\rangle\}))\}.$$

Proposition 6.16 *For an interval-valued intuitionistic fuzzy soft multi-set (F, A) over U,*

(a) $((F, A)^c)^c = (F, A)$,
(b) $(F, A)^c_{\approx \Phi_i} = (F, A)_{\approx U_i}$,
(c) $(F, A)^c_\Phi = (F, A)_U$,
(d) $(F, A)^c_{\approx U_i} = (F, A)_{\approx \Phi_i}$,
(e) $(F, A)^c_U = (F, A)_\Phi$,

Proof The proof is straightforward.   □

Definition 6.17 The *union* of two interval-valued intuitionistic fuzzy soft multi-sets (F, A) and (G, B) over U is an interval-valued intuitionistic fuzzy soft multi-set (H, D) where

$$D = A \cup B \quad \text{and} \quad \forall e \in D,$$

$$H(e) = \begin{cases} F(e), & \text{if } e \in A - B \\ G(e), & \text{if } e \in B - A \\ \bigcup (F(e), G(e)), & \text{if } e \in A \cap B \end{cases}$$

where $\bigcup (F(e), G(e)) = F_{e_{U_{i,j}}} \cup\% G_{e_{U_{i,j}}} \forall i \in \{1, 2, 3, \ldots, m\}$ with $\cup\%$ as an interval-valued intuitionistic fuzzy union and is written as $(F, A) \tilde{\cup} (G, B) = (H, D)$.

Example 6.18 Let us consider there are three universes U_1, U_2, and U_3. Let $U_1 = \{h_1, h_2, h_3\}$, $U_2 = \{c_1, c_2, c_3\}$, and $U_3 = \{v_1, v_2\}$. Let $\{E_{U_1}, E_{U_2}, E_{U_3}\}$ be a collection of sets of decision parameters related to the above universes where

$$E_{U_1} = \{e_{U_1,1}, e_{U_1,2}, e_{U_1,3}\}, \quad E_{U_2} = \{e_{U_2,1}, e_{U_2,2}, e_{U_2,3}\},$$
$$E_{U_3} = \{e_{U_3,1}, e_{U_3,2}, e_{U_3,3}\}.$$

Let

$$A = \{a_1 = (e_{U_1,1}, e_{U_2,1}, e_{U_3,1}), a_2 = (e_{U_1,1}, e_{U_2,2}, e_{U_3,1}), a_3 = (e_{U_1,2}, e_{U_2,2}, e_{U_3,1})\},$$
$$B = \{b_1 = (e_{U_1,1}, e_{U_2,1}, e_{U_3,1}), b_2 = (e_{U_1,1}, e_{U_2,2}, e_{U_3,1}), b_3 = (e_{U_1,2}, e_{U_2,3}, e_{U_3,2})\}$$

Let (F, A) and (G, B) be two interval-valued intuitionistic fuzzy soft multi-set over U, such that

$$(F, A) = \{(a_1, (\{\langle h_1, [0.1, 0.3], [0.3, 0.5]\rangle, \langle h_2, [0.3, 0.4], [0.4, 0.6]\rangle,$$
$$\langle h_3, [0.7, 0.9], [0, 0.1]\rangle\}, \{\langle c_1, [0.7, 0.9], [0, 0]\rangle,$$
$$\langle c_2, [0.4, 0.6], [0.3, 0.4]\rangle, \langle c_3, [0.3, 0.5], [0.2, 0.4]\rangle\},$$
$$\{\langle v_1, [0.7, 0.9], [0, 0]\rangle, \langle v_2, [0.6, 0.8], [0, 0.2]\rangle\})),$$
$$(a_2, (\{\langle h_1, [0.1, 0.3], [0.3, 0.5]\rangle, \langle h_2, [0.3, 0.5], [0.4, 0.5]\rangle,$$
$$\langle h_3, [0.7, 0.9], [0, 0.1]\rangle\}, \{\langle c_1, [0.3, 0.5], [0.2, 0.4]\rangle,$$
$$\langle c_2, [0.4, 0.6], [0.3, 0.4]\rangle, \langle c_3, [0.7, 0.9], [0, 0]\rangle\},$$
$$\{\langle v_1, [0.3, 0.5], [0.2, 0.4]\rangle, \langle v_2, [0.3, 0.5], [0.2, 0.4]\rangle\})),$$
$$(a_3, (\{\langle h_1, [0.6, 0.8], [0.1, 0.2]\rangle, \langle h_2, [0.6, 0.8], [0.1, 0.2]\rangle,$$
$$\langle h_3, [0, 0.2], [0.5, 0.7]\rangle\}, \{\langle c_1, [0.7, 0.9], [0, 0.1]\rangle,$$
$$\langle c_2, [0.5, 0.7], [0.1, 0.3]\rangle, \langle c_3, [0.2, 0.4], [0.3, 0.5]\rangle\},$$
$$\{\langle v_1, [0.4, 0.6], [0.2, 0.4]\rangle, \langle v_2, [0.3, 0.5], [0.2, 0.4]\rangle\}))\}.$$

$$(G, B) = \{(b_1, (\{\langle h_1, [0.2, 0.4], [0.3, 0.5]\rangle, \langle h_2, [0.4, 0.6], [0.2, 0.4]\rangle,$$
$$\langle h_3, [0.7, 0.9], [0, 0.1]\rangle\}, \{\langle c_1, [0.7, 0.9], [0, 0]\rangle,$$
$$\langle c_2, [0.6, 0.8], [0.1, 0.2]\rangle, \langle c_3, [0.4, 0.6], [0.2, 0.4]\rangle\},$$
$$\{\langle v_1, [0.7, 0.9], [0, 0]\rangle, \langle v_2, [0.7, 0.9], [0, 0.1]\rangle\})),$$
$$(b_2, (\{\langle h_1, [0.3, 0.5], [0.2, 0.3]\rangle, \langle h_2, [0.3, 0.5], [0.2, 0.4]\rangle,$$
$$\langle h_3, [0.7, 0.9], [0, 0.1]\rangle\}, \{\langle c_1, [0.3, 0.5], [0.2, 0.4]\rangle,$$
$$\langle c_2, [0.4, 0.6], [0.3, 0.4]\rangle, \langle c_3, [0.7, 0.9], [0, 0]\rangle\},$$
$$\{\langle v_1, [0.5, 0.7], [0.2, 0.3]\rangle, \langle v_2, [0.3, 0.5], [0.2, 0.4]\rangle\})),$$
$$(b_3, (\{\langle h_1, [0.6, 0.8], [0.1, 0.2]\rangle, \langle h_2, [0.6, 0.8], [0.1, 0.2]\rangle,$$
$$\langle h_3, [0, 0.2], [0.5, 0.7]\rangle\}, \{\langle c_1, [0.7, 0.9], [0, 0.1]\rangle,$$
$$\langle c_2, [0.5, 0.7], [0.1, 0.3]\rangle, \langle c_3, [0.2, 0.4], [0.3, 0.5]\rangle\},$$
$$\{\langle v_1, [0.4, 0.6], [0.2, 0.4]\rangle, \langle v_2, [0.3, 0.5], [0.2, 0.4]\rangle\})).$$

Then,

$$(F,A) \, \tilde{\cup} \, (G,B) = (H,D) = \{(d_1,(\{\langle h_1,[0.2,0.4],[0.3,0.5]\rangle, \langle h_2,[0.4,0.6],[0.2,0.4]\rangle,$$
$$\langle h_3,[0.7,0.9],[0,0.1]\rangle\}, \{\langle c_1,[0.7,0.9],[0,0]\rangle, \langle c_2,[0.6,0.8],[0.1,0.2]\rangle,$$
$$\langle c_3,[0.4,0.6],[0.2,0.4]\rangle\}, \{\langle v_1,[0.7,0.9],[0,0]\rangle, \langle v_2,[0.7,0.9],[0,0.1]\rangle\})),$$
$$(d_2,(\{\langle h_1,[0.3,0.5],[0.2,0.3]\rangle, \langle h_2,[0.3,0.5],[0.2,0.4]\rangle,$$
$$\langle h_3,[0.7,0.9],[0,0.1]\rangle\}, \{\langle c_1,[0.3,0.5],[0.2,0.4]\rangle, \langle c_2,[0.4,0.6],[0.3,0.4]\rangle,$$
$$\langle c_3,[0.7,0.9],[0,0]\rangle\}, \{\langle v_1,[0.5,0.7],[0.2,0.3]\rangle, \langle v_2,[0.3,0.5],[0.2,0.4]\rangle\})),$$
$$(d_3,(\{\langle h_1,[0.6,0.8],[0.1,0.2]\rangle, \langle h_2,[0.6,0.8],[0.1,0.2]\rangle,$$
$$\langle h_3,[0,0.2],[0.5,0.7]\rangle\}, \{\langle c_1,[0.7,0.9],[0,0.1]\rangle, \langle c_2,[0.5,0.7],[0.1,0.3]\rangle,$$
$$\langle c_3,[0.2,0.4],[0.3,0.5]\rangle\}\{\langle v_1,[0.4,0.6],[0.2,0.4]\rangle, \langle v_2,[0.3,0.5],[0.2,0.4]\rangle\})),$$
$$(d_4,(\{\langle h_1,[0.6,0.8],[0.1,0.2]\rangle, \langle h_2,[0.6,0.8],[0.1,0.2]\rangle,$$
$$\langle h_3,[0,0.2],[0.5,0.7]\rangle\}, \{\langle c_1,[0.7,0.9],[0,0.1]\rangle, \langle c_2,[0.5,0.7],[0.1,0.3]\rangle,$$
$$\langle c_3,[0.2,0.4],[0.3,0.5]\rangle\}, \{\langle v_1,[0.4,0.6],[0.2,0.4]\rangle, \langle v_2,[0.3,0.5],[0.2,0.4]\rangle\}))\},$$

where $D = \{d_1 = a_1 = b_1, d_2 = a_2 = b_2, d_3 = a_3, d_4 = b_3\}$.

Proposition 6.19 *If (F, A), (G, B), and (H, C) are three interval-valued intuitionistic fuzzy soft multi-sets over U, then*

(a) $(F,A) \, \tilde{\cup} \, ((G,B) \, \tilde{\cup} \, (H,C)) = ((F,A) \, \tilde{\cup} \, (G,B)) \, \tilde{\cup} \, (H,C),$

(b) $(F,A) \, \tilde{\cup} \, (F,A) = (F,A),$

(c) $(F,A) \, \tilde{\cup} \, (G,A)_{\approx \Phi_i} = (R,A)$, where R is defined by (3.16)

(d) $(F,A) \, \tilde{\cup} \, (G,A)_{\Phi} = (F,A),$

(e) $(F,A) \, \tilde{\cup} \, (G,B)_{\approx \Phi_i} = (R,D)$, where $D = A \cup B$ and R is defined by (3.16)

(f) $(F,A) \, \tilde{\cup} \, (G,B)_{\Phi} = \begin{cases} (F,A), & \text{if } A = B \\ (R,D), & \text{otherwise} \end{cases}$ where $D = A \cup B$

(g) $(F,A) \, \tilde{\cup} \, (G,A)_{U_i} = (R,A)_{\approx U_i},$

(h) $(F,A) \, \tilde{\cup} \, (G,A)_U = (G,A)_U,$

(i) $(F,A) \, \tilde{\cup} \, (G,B)_{U_i} = \begin{cases} (R,D)_{U_i}, & \text{if } A = B \\ (R,D), & \text{otherwise} \end{cases}$ where $D = A \cup B$

(j) $(F,A) \, \tilde{\cup} \, (G,B)_U = \begin{cases} (G,B)_U, & \text{if } A = B \\ (R,D), & \text{otherwise} \end{cases}$ where $D = A \cup B$

Proof The proof is straight forward. □

Definition 6.20 The *intersection* of two interval-valued intuitionistic fuzzy soft multi-sets (F, A) and (G, B) over U is an interval-valued intuitionistic fuzzy soft multi-set (H, D) where

$$D = A \cap B \quad \text{and} \quad \forall e \in D,$$

$$H(e) = \begin{cases} F(e), & \text{if } e \in A - B \\ G(e), & \text{if } e \in B - A \\ \cap(F(e), G(e)), & \text{if } e \in A \cap B \end{cases}$$

where $\cap(F(e), G(e)) = F_{e_{U_{i,j}}} \cap \% \, G_{e_{U_{i,j}}} \, \forall i \in \{1, 2, 3, \ldots, n\}$ with $\cap \%$ as an interval-valued intuitionistic fuzzy intersection and is written as $(F, A) \tilde{\cap} (G, B) = (H, C)$.

Example 6.21 Let us consider Example 6.18. Then,

$(F, A) \tilde{\cap} (G, B) = (H, D) = \{(d_1, (\{\langle h_1, [0.1, 0.3], [0.3, 0.5]\rangle, \langle h_2, [0.3, 0.4], [0.4, 0.6]\rangle,$
$\langle h_3, [0.7, 0.9], [0, 0.1]\rangle\}, \{\langle c_1, [0.7, 0.9], [0, 0]\rangle, \langle c_2, [0.4, 0.6], [0.3, 0.4]\rangle,$
$\langle c_3, [0.3, 0.5], [0.2, 0.4]\rangle\}, \{\langle v_1, [0.7, 0.9], [0, 0]\rangle, \langle v_2, [0.6, 0.8], [0, 0.2]\rangle\})),$
$(d_2, (\{\langle h_1, [0.1, 0.3], [0.3, 0.5]\rangle, \langle h_2, [0.3, 0.5], [0.4, 0.5]\rangle,$
$\langle h_3, [0.7, 0.9], [0, 0.1]\rangle\}, \{\langle c_1, [0.3, 0.5], [0.2, 0.4]\rangle, \langle c_2, [0.4, 0.6], [0.3, 0.4]\rangle,$
$\langle c_3, [0.7, 0.9], [0, 0]\rangle\}, \{\langle v_1, [0.3, 0.5], [0.2, 0.4]\rangle, \langle v_2, [0.3, 0.5], [0.2, 0.4]\rangle\})),$
$(d_3, (\{\langle h_1, [0.6, 0.8], [0.1, 0.2]\rangle, \langle h_2, [0.6, 0.8], [0.1, 0.2]\rangle,$
$\langle h_3, [0, 0.2], [0.5, 0.7]\rangle\}, \{\langle c_1, [0.7, 0.9], [0, 0.1]\rangle, \langle c_2, [0.5, 0.7], [0.1, 0.3]\rangle,$
$\langle c_3, [0.2, 0.4], [0.3, 0.5]\rangle\}, \{\langle v_1, [0.4, 0.6], [0.2, 0.4]\rangle, \langle v_2, [0.3, 0.5], [0.2, 0.4]\rangle\})),$
$(d_4, (\{\langle h_1, [0.6, 0.8], [0.1, 0.2]\rangle, \langle h_2, [0.6, 0.8], [0.1, 0.2]\rangle,$
$\langle h_3, [0, 0.2], [0.5, 0.7]\rangle\}, \{\langle c_1, [0.7, 0.9], [0, 0.1]\rangle, \langle c_2, [0.5, 0.7], [0.1, 0.3]\rangle,$
$\langle c_3, [0.2, 0.4], [0.3, 0.5]\rangle\}, \{\langle v_1, [0.4, 0.6], [0.2, 0.4]\rangle, \langle v_2, [0.3, 0.5], [0.2, 0.4]\rangle\}))\},$

where $D = \{d_1 = a_1 = b_1, d_2 = a_2 = b_2, d_3 = a_3, d_4 = b_3\}$.

Proposition 6.22 *If (F, A), (G, B), and (H, C) are three interval-valued intuitionistic fuzzy soft multi-sets over U, then*

(a) $(F, A) \tilde{\cap} ((G, B) \tilde{\cap} (H, C)) = ((F, A) \tilde{\cap} (G, B)) \tilde{\cap} (H, C)$,

(b) $(F, A) \tilde{\cap} (F, A) = (F, A)$,

(c) $(F, A) \tilde{\cap} (G, A)_{\approx \Phi_i} = (R, A)_{\approx \Phi_i}$, where R is defined by (5.19)

(d) $(F, A) \tilde{\cap} (G, A)_{\Phi} = (R, A)_{\Phi}$,

(e) $(F, A) \tilde{\cap} (G, B)_{\approx \Phi_i} = \begin{cases} (R, D)_{\Phi_i}, & \text{if } A \subseteq B \\ (R, D), & \text{otherwise} \end{cases}$ where $D = A \cup B$, where $D =$
   $A \cup B$ and R is defined by (5.19),

(f) $(F, A) \tilde{\cap} (G, B)_{\Phi} = \begin{cases} (R, D)_{\Phi}, & \text{if } A \subseteq B \\ (R, D), & \text{otherwise} \end{cases}$ where $D = A \cup B$ and R is defined
   by (3.19),

(g) $(F, A) \tilde{\cap} (G, A)_{U_i} = (R, D)$, where $D = A \cup B$ and R is defined by (5.19),

(h) $(F,A) \mathbin{\tilde{\cap}} (G,A)_U = (F,A),$

(i) $(F,A) \mathbin{\tilde{\cap}} (G,B)_{U_i} = (R,D)$ where $D = A \cup B$ and R is defined by (3.19),

(j) $(F,A) \mathbin{\tilde{\cap}} (G,B)_U = \begin{cases} (F,A), & \text{if } A \supseteq B \\ (R,D), & \text{otherwise} \end{cases}$ where, $D = A \cup B$ and R is defined

by (3.19).

Proof The proof is straight forward. $\square$

6.1 Relations on Interval-Valued Intuitionistic Fuzzy Soft Multi-Sets

The concept of interval-valued intuitionistic fuzzy soft multi-set relations (IVIFSMS-relations) is given in this section. Also the basic properties of the IVIFSMS-relations and various types of IVIFSMS-relations are presented in this section.

Definition 6.23 Let $\{U_i : i \in I\}$ be a collection of universes such that $\bigcap_{i \in I} U_i = \phi$ and let $\{E_{U_i} : i \in I\}$ be a collection of sets of parameters. Let $U = \prod_{i \in I} \text{IVIFS}(U_i)$ where $\text{IVIFS}(U_i)$ denotes the set of all interval-valued intuitionistic fuzzy subsets of U_i, $E = \prod_{i \in I} E_{U_i}$, and $A, B \subseteq E$. Let (F, A) and (G, B) be two interval-valued intuitionistic fuzzy soft multi-sets over U where F and G are mappings given by $F, G : A \to U$. Then, a relation R between them is defined as a pair $(H, A \times B)$ where H is mapping given by $H : A \times B \to U$. The collection of relations on interval-valued intuitionistic fuzzy soft multi-sets on $A \times B$ over U is denoted by $\text{MSSR}_U(A \times B)$.

Example 6.24 Let us consider there are three universes U_1, U_2, and U_3. Let $U_1 = \{h_1, h_2, h_3\}$, $U_2 = \{c_1, c_2\}$, and $U_3 = \{v_1, v_2\}$. Let $\{E_{U_1}, E_{U_2}, E_{U_3}\}$ be a collection of sets of decision parameters related to the above universes where

$$E_{U_1} = \{e_{U_1,1}, e_{U_1,2}, e_{U_1,3}\}, \quad E_{U_2} = \{e_{U_2,1}, e_{U_2,2}\}, \quad E_{U_3} = \{e_{U_3,1}, e_{U_3,2}\}$$

Let $U = \prod_{i=1}^{3} \text{IVIFS}(U_i)$, $E = \prod_{i=1}^{3} E_{U_i}$, and $A, B \subseteq E$, such that

$$A = \{a_1 = (e_{U_1,1}, e_{U_2,1}, e_{U_3,1}), a_2 = (e_{U_1,1}, e_{U_2,2}, e_{U_3,1})\} \text{ and}$$
$$B = \{b_1 = (e_{U_1,2}, e_{U_2,2}, e_{U_3,1}), b_2 = (e_{U_1,1}, e_{U_2,2}, e_{U_3,2})\}.$$

Let the tabular representation of the interval-valued intuitionistic fuzzy soft multi-set (F, A) be

	a_1	a_2
h_1	[0.1, 0.3], [0.3, 0.5]	[0.2, 0.5], [0.3, 0.4]
h_2	[0.3, 0.4], [0.4, 0.6]	[0.1, 0.2], [0.4, 0.7]
h_3	[0.7, 0.9], [0.0, 0.1]	[0.3, 0.6], [0.2, 0.4]
c_1	[0.7, 0.9], [0.0, 0.0]	[0.3, 0.6], [0.2, 0.4]
c_2	[0.4, 0.6], [0.3, 0.4]	[0.4, 0.5], [0.1, 0.3]
v_1	[0.7, 0.9], [0.0, 0.0]	[0.6, 0.8], [0.0, 0.0]
v_2	[0.6, 0.8], [0.1, 0.2]	[0.4, 0.7], [0.1, 0.2]

Let the tabular representation of the interval-valued intuitionistic fuzzy soft multi-set (G, B) is

	b_1	b_2
h_1	[0.2, 0.3], [0.4, 0.6]	[0.1, 0.2], [0.2, 0.4]
h_2	[0.4, 0.5], [0.3, 0.4]	[0.3, 0.4], [0.1, 0.3]
h_3	[0.6, 0.7], [0.2, 0.3]	[0.4, 0.6], [0.2, 0.3]
c_1	[0.8, 0.9], [0.0, 0.1]	[0.5, 0.8], [0.0, 0.0]
c_2	[0.3, 0.5], [0.2, 0.4]	[0.3, 0.7], [0.1, 0.2]
v_1	[0.4, 0.7], [0.2, 0.3]	[0.2, 0.4], [0.2, 0.5]
v_2	[0.5, 0.7], [0.1, 0.3]	[0.6, 0.8], [0.1, 0.2]

Then, a relation $R_1(=(H, A \times B)$, say) between them is given by

	(a_1, b_1)	(a_1, b_2)	(a_2, b_1)	(a_2, b_2)
h_1	[0.2, 0.3], [0.4, 0.6]	[0.6, 0.8], [0.1, 0.2]	[0.2, 0.4], [0.2, 0.4]	[0.5, 0.7], [0.1, 0.3]
h_2	[0.4, 0.5], [0.3, 0.4]	[0.4, 0.7], [0.2, 0.3]	[0.6, 0.7], [0.1, 0.2]	[0.4, 0.5], [0.3, 0.4]
h_3	[0.4, 0.6], [0.3, 0.4]	[0.2, 0.4], [0.2, 0.5]	[0.2, 0.3], [0.4, 0.6]	[0.2, 0.4], [0.4, 0.6]
c_1	[0.2, 0.5], [0.1, 0.3]	[0.1, 0.4], [0.4, 0.6]	[0.5, 0.7], [0.1, 0.3]	[0.3, 0.6], [0.1, 0.3]
c_2	[0.5, 0.7], [0.2, 0.3]	[0.4, 0.5], [0.3, 0.4]	[0.1, 0.3], [0.4, 0.5]	[0.1, 0.2], [0.4, 0.7]
v_1	[0.3, 0.6], [0.1, 0.3]	[0.3, 0.7], [0.1, 0.2]	[0.6, 0.8], [0.1, 0.2]	[0.4, 0.7], [0.2, 0.3]
v_2	[0.6, 0.8], [0.1, 0.2]	[0.2, 0.3], [0.4, 0.6]	[0.2, 0.5], [0.3, 0.4]	[0.2, 0.3], [0.4, 0.5]

Let a relation R_2 (=$(J, A \times B)$, say) between them is given by

	(a_1, b_1)	(a_1, b_2)	(a_2, b_1)	(a_2, b_2)
h_1	[0.4, 0.6], [0.3, 0.4]	[0.4, 0.5], [0.2, 0.4]	[0.3, 0.4], [0.1, 0.3]	[0.4, 0.7], [0.1, 0.2]
h_2	[0.1, 0.5], [0.1, 0.4]	[0.2, 0.3], [0.6, 0.7]	[0.5, 0.6], [0.2, 0.3]	[0.4, 0.5], [0.3, 0.4]
h_3	[0.3, 0.4], [0.3, 0.5]	[0.4, 0.7], [0.1, 0.3]	[0.2, 0.3], [0.4, 0.6]	[0.2, 0.4], [0.4, 0.6]
c_1	[0.2, 0.3], [0.2, 0.4]	[0.1, 0.2], [0.4, 0.6]	[0.3, 0.6], [0.2, 0.4]	[0.4, 0.6], [0.1, 0.4]
c_2	[0.5, 0.6], [0.1, 0.3]	[0.3, 0.6], [0.2, 0.4]	[0.4, 0.5], [0.2, 0.3]	[0.1, 0.2], [0.4, 0.7]
v_1	[0.4, 0.5], [0.2, 0.3]	[0.5, 0.7], [0.1, 0.3]	[0.4, 0.6], [0.3, 0.4]	[0.3, 0.7], [0.1, 0.2]
v_2	[0.0, 0.1], [0.7, 0.9]	[0.2, 0.3], [0.6, 0.7]	[0.2, 0.4], [0.5, 0.6]	[0.2, 0.3], [0.3, 0.6]

The tabular representations of R_1 and R_2 are called *relational matrices* for R_1 and R_2, respectively.

From above, we have $\mu_{H(a_1,b_2)}(h_1) = [0.6, \ 0.8]$ and $\gamma_{J(a_1,b_2)}(c_2) = [0.2, 0.4]$, etc. But these intervals lie on the 1st row–2nd column and 5th row–2nd column respectively. So we denote $\mu_{H(e_1,e_2)}(h_1)\big|_{(1,2)} = [0.6, 0.8]$ and $\gamma_{J(e_1,e_2)}(h_2)$ $\big|_{(5,2)} = [0.2, 0.4]$, etc., to make the clear concept about what are the positions of the intervals in the relational matrices.

Remark 6.25 Let (F_1, A_1), (F_2, A_2), …, (F_n, A_n) be n numbers of interval-valued intuitionistic fuzzy soft multi-sets over U. Then, a relation R between them is defined as a pair $(H, A_1 \times A_2 \times \cdots \times A_n)$ where H is mapping given by $H: A_1 \times A_2 \times \cdots \times A_n \to U$.

Definition 6.26 Let $\{U_i : i \in I\}$ be a collection of universes such that $\bigcap_{i \in I} U_i = \phi$ and let $\{E_{U_i} : i \in I\}$ be a collection of sets of parameters. Let $U = \prod_{i \in I} \text{IVIFS}(U_i)$ where $\text{IVIFS}(U_i)$ denotes the set of all interval-valued intuitionistic fuzzy subsets of U_i, $E = \prod_{i \in I} E_{U_i}$ and $A, B \subseteq E$. Let (F, A) and (G, B) be two interval-valued intuitionistic fuzzy soft multi-sets over U where F and G are mappings given by $F, G: A \to U$. Let R be a relation between them. Then, the *order* of the relational matrix is (α, β) where $\alpha = \Sigma_i \ n(U_i)$ and β = number of pairs of parameters considered in the relational matrix.

In Example 6.24, both the relational matrices for R_1 and R_2 are of order (7, 4). If $\alpha = \beta$, then the relation matrix is called a square matrix.

Definition 6.27 Let $R_1, R_2 \in \text{MSSR}_U(A \times B)$, $R_1 = (H, A \times B)$, $R_2 = (J, A \times B)$. Then, we define

(i) $R_1 \vee R_2 = (H \blacklozenge J, A \times B)$ where $H \blacklozenge J: A \times B \to U$ is defined as $(H \blacklozenge J)(a_i, b_j) = H(a_i, b_j) \cup\% \ J(a_i, b_j)$ for $(a_i, b_j) \in A \times B$ where $\cup\%$ denotes the interval-valued intuitionistic fuzzy union.

(ii) $R_1 \wedge R_2 = (H \bullet J, A \times B)$ where $H \bullet J: A \times B \to U$ is defined as $(H \bullet J)(a_i, b_j) = H(a_i, b_j) \cap\% \ J(a_i, b_j)$ for $(a_i, b_j) \in A \times B$ where $\cap\%$ denotes the interval-valued intuitionistic fuzzy intersection.

(iii) $R_1^c = (\sim H, A \times B)$ where $\sim H: A \times B \rightarrow U$ is defined as
$\sim H(a_i, b_j) = [H(a_i, b_j)]^{\#}$ for $(a_i, b_j) \in A \times B$ where # denotes the interval-valued intuitionistic fuzzy complement.

Example 6.28 Consider Example 6.24. Then, we get
$R_1 \vee R_2$:

	(a_1, b_1)	(a_1, b_2)	(a_2, b_1)	(a_2, b_2)
h_1	([0.4, 0.6], [0.3, 0.4])	([0.6, 0.8], [0.1, 0.2])	([0.3, 0.4], [0.1, 0.3])	([0.5, 0.7], [0.1, 0.2])
h_2	([0.4, 0.5], [0.1, 0.4])	([0.4, 0.7], [0.2, 0.3])	([0.6, 0.7], [0.1, 0.2])	([0.4, 0.5], [0.3, 0.4])
h_3	([0.4, 0.6], [0.3, 0.4])	([0.4, 0.7], [0.1, 0.3])	([0.2, 0.3], [0.4, 0.6])	([0.2, 0.4], [0.4, 0.6])
c_1	([0.2, 0.5], [0.1, 0.3])	([0.1, 0.4], [0.4, 0.6])	([0.5, 0.7], [0.1, 0.3])	([0.4, 0.6], [0.1, 0.3])
c_2	([0.5, 0.7], [0.1, 0.3])	([0.4, 0.6], [0.2, 0.4])	([0.4, 0.5], [0.2, 0.3])	([0.1, 0.2], [0.4, 0.7])
v_1	([0.4, 0.6], [0.1, 0.3])	([0.5, 0.7], [0.1, 0.2])	([0.6, 0.8], [0.1, 0.2])	([0.4, 0.7], [0.1, 0.2])
v_2	([0.6, 0.8], [0.1, 0.2])	([0.2, 0.3], [0.4, 0.6])	([0.2, 0.5], [0.3, 0.4])	([0.2, 0.3], [0.3, 0.5])

$R_1 \wedge R_2$:

	(a_1, b_1)	(a_1, b_2)	(a_2, b_1)	(a_2, b_2)
h_1	([0.2, 0.3], [0.4, 0.6])	([0.4, 0.5], [0.2, 0.4])	([0.2, 0.4], [0.2, 0.4])	([0.4, 0.7], [0.1, 0.3])
h_2	([0.1, 0.5], [0.3, 0.4])	([0.2, 0.3], [0.6, 0.7])	([0.5, 0.6], [0.2, 0.3])	([0.4, 0.5], [0.3, 0.4])
h_3	([0.3, 0.4], [0.3, 0.5])	([0.2, 0.4], [0.2, 0.5])	([0.2, 0.3], [0.4, 0.6])	([0.2, 0.4], [0.4, 0.6])
c_1	([0.2, 0.3], [0.2, 0.4])	([0.1, 0.2], [0.4, 0.6])	([0.3, 0.6], [0.2, 0.4])	([0.3, 0.6], [0.1, 0.4])
c_2	([0.5, 0.6], [0.2, 0.3])	([0.3, 0.5], [0.3, 0.4])	([0.1, 0.3], [0.4, 0.5])	([0.1, 0.2], [0.4, 0.7])
v_1	([0.3, 0.5], [0.2, 0.3])	([0.3, 0.7], [0.1, 0.3])	([0.4, 0.6], [0.3, 0.4])	([0.3, 0.7], [0.2, 0.3])
v_2	([0.0, 0.1], [0.7, 0.9])	([0.2, 0.3], [0.6, 0.7])	([0.2, 0.4], [0.5, 0.6])	([0.2, 0.3], [0.4, 0.5])

R_1^c:

	$(a_1, b_1)^c$	$(a_1, b_1)^c$	$(a_1, b_1)^c$	$(a_1, b_1)^c$
h_1	([0.4, 0.6], [0.2, 0.3])	([0.1, 0.2], [0.6, 0.8])	([0.2, 0.4], [0.2, 0.4])	([0.1, 0.3], [0.5, 0.7])
h_2	([0.3, 0.4], [0.4, 0.5])	([0.2, 0.3], [0.4, 0.7])	([0.1, 0.2], [0.6, 0.7])	([0.3, 0.4], [0.4, 0.5])
h_3	([0.3, 0.4], [0.4, 0.6])	([0.2, 0.5], [0.2, 0.4])	([0.4, 0.6], [0.2, 0.3])	([0.4, 0.6], [0.2, 0.4])
c_1	([0.1, 0.3], [0.2, 0.5])	([0.4, 0.6], [0.1, 0.4])	([0.1, 0.3], [0.5, 0.7])	([0.1, 0.3], [0.3, 0.6])
c_2	([0.2, 0.3], [0.5, 0.7])	([0.3, 0.4], [0.4, 0.5])	([0.4, 0.5], [0.1, 0.3])	([0.4, 0.7], [0.1, 0.2])
v_1	([0.1, 0.3], [0.3, 0.6])	([0.1, 0.2], [0.3, 0.7])	([0.1, 0.2], [0.6, 0.8])	([0.2, 0.3], [0.4, 0.7])
v_2	([0.1, 0.2], [0.6, 0.8])	([0.4, 0.6], [0.2, 0.3])	([0.3, 0.4], [0.2, 0.5])	([0.4, 0.5], [0.2, 0.3])

Result 6.29 Let $R_1, R_2, R_3 \in \text{MSSR}_U(A \times B)$. Then, the following properties hold:

(a) $(R_1 \vee R_2)^c = R_1^c \wedge R_2^c$,
(b) $(R_1 \wedge R_2)^c = R_1^c \vee R_2^c$,
(c) $R_1 \vee (R_2 \vee R_3) = (R_1 \vee R_2) \vee R_3$,
(d) $R_1 \wedge (R_2 \wedge R_3) = (R_1 \wedge R_2) \wedge R_3$,
(e) $R_1 \wedge (R_2 \vee R_3) = (R_1 \wedge R_2) \vee (R_1 \wedge R_3)$, and
(f) $R_1 \vee (R_2 \wedge R_3) = (R_1 \vee R_2) \wedge (R_1 \vee R_3)$.

Definition 6.30 Let $R_1, R_2 \in \text{MSSR}_U(A \times B)$. Then, $R_1 \leq R_2$ iff $H(a_i, b_j) \subseteq J(a_i, b_j)$ for $(a_i, b_j) \in A \times B$ where $R_1 = (H, A \times B)$ and $R_2 = (J, A \times B)$.

Example 6.31 Consider the interval-valued intuitionistic fuzzy soft multi-sets (F, A) and (G, B) given in 6.24. Let $R_1, R_2 \in \text{MSSR}_U(A \times B)$ be defined as follows:
R_1:

	(a_1, b_1)	(a_1, b_2)	(a_2, b_1)	(a_2, b_2)
h_1	[0.2, 0.3], [0.4, 0.6]	[0.6, 0.8], [0.1, 0.2]	[0.2, 0.4], [0.2, 0.4]	[0.5, 0.7], [0.1, 0.3]
h_2	[0.4, 0.5], [0.3, 0.4]	[0.4, 0.7], [0.2, 0.3]	[0.6, 0.7], [0.1, 0.2]	[0.4, 0.5], [0.3, 0.4]
h_3	[0.4, 0.6], [0.3, 0.4]	[0.2, 0.4], [0.2, 0.5]	[0.2, 0.3], [0.4, 0.6]	[0.2, 0.4], [0.4, 0.6]
c_1	[0.2, 0.5], [0.1, 0.3]	[0.1, 0.4], [0.4, 0.6]	[0.5, 0.7], [0.1, 0.3]	[0.3, 0.6], [0.1, 0.3]
c_2	[0.5, 0.7], [0.2, 0.3]	[0.4, 0.5], [0.3, 0.4]	[0.1, 0.3], [0.4, 0.5]	[0.1, 0.2], [0.4, 0.7]
v_1	[0.3, 0.6], [0.1, 0.3]	[0.3, 0.7], [0.1, 0.3]	[0.6, 0.8], [0.1, 0.2]	[0.4, 0.7], [0.2, 0.3]
v_2	[0.6, 0.8], [0.1, 0.2]	[0.2, 0.3], [0.4, 0.6]	[0.2, 0.5], [0.3, 0.4]	[0.2, 0.3], [0.4, 0.5]

R_2:

	(a_1, b_1)	(a_1, b_2)	(a_2, b_1)	(a_2, b_2)
h_1	[0.2, 0.4], [0.3, 0.4]	[0.6, 0.8], [0.1, 0.2]	[0.3, 0.5], [0.1, 0.3]	[0.5, 0.8], [0.1, 0.2]
h_2	[0.4, 0.6], [0.1, 0.4]	[0.6, 0.7], [0.2, 0.3]	[0.6, 0.8], [0.1, 0.2]	[0.3, 0.6], [0.3, 0.4]
h_3	[0.4, 0.6], [0.3, 0.4]	[0.2, 0.4], [0.2, 0.4]	[0.2, 0.4], [0.4, 0.5]	[0.2, 0.5], [0.4, 0.5]
c_1	[0.2, 0.5], [0.1, 0.2]	[0.1, 0.5], [0.3, 0.4]	[0.5, 0.7], [0.1, 0.2]	[0.3, 0.6], [0.1, 0.2]
c_2	[0.6, 0.7], [0.1, 0.2]	[0.4, 0.5], [0.3, 0.4]	[0.2, 0.4], [0.4, 0.5]	[0.2, 0.3], [0.3, 0.6]
v_1	[0.4, 0.8], [0.1, 0.2]	[0.3, 0.8], [0.1, 0.2]	[0.6, 0.8], [0.1, 0.2]	[0.4, 0.7], [0.2, 0.3]
v_2	[0.6, 0.9], [0.0, 0.1]	[0.2, 0.3], [0.3, 0.5]	[0.5, 0.6], [0.2, 0.3]	[0.3, 0.4], [0.4, 0.5]

Then, clearly $R_1 \leq R_2$.

Definition 6.32 Let $\{U_i : i \in I\}$ be a collection of universes such that $\bigcap_{i \in I} U_i = \phi$ and let $\{E_{U_i} : i \in I\}$ be a collection of sets of parameters. Let $U = \prod_{i \in I} \text{IVIFS}(U_i)$ where $\text{IVIFS}(U_i)$ denotes the set of all interval-valued intuitionistic fuzzy subsets of U_i, $E = \prod_{i \in I} E_{U_i}$ and $A, B \subseteq E$. Let (F, A) and (G, B) be two interval-valued intuitionistic fuzzy soft multi-sets over U where F and G are mappings given by $F, G : A \rightarrow U$. Then,

(i) a *null relation* O_U between them is defined as

$$O_U = (H, A \times B)_\phi$$

(ii) an *absolute relation* I_U between them is defined as

$$I_U = (H, A \times B)_U.$$

Example 6.33 Consider the interval-valued intuitionistic fuzzy soft multi-sets (F, A) and (G, B) given in 6.24. Then, a *null relation* O_U between them is given by O_U:

	(a_1, b_1)	(a_1, b_2)	(a_2, b_1)	(a_2, b_2)
h_1	[0.0, 0.0], [1.0, 1.0]	[0.0, 0.0], [1.0, 1.0]	[0.0, 0.0], [1.0, 1.0]	[0.0, 0.0], [1.0, 1.0]
h_2	[0.0, 0.0], [1.0, 1.0]	[0.0, 0.0], [1.0, 1.0]	[0.0, 0.0], [1.0, 1.0]	[0.0, 0.0], [1.0, 1.0]
h_3	[0.0, 0.0], [1.0, 1.0]	[0.0, 0.0], [1.0, 1.0]	[0.0, 0.0], [1.0, 1.0]	[0.0, 0.0], [1.0, 1.0]
c_1	[0.0, 0.0], [1.0, 1.0]	[0.0, 0.0], [1.0, 1.0]	[0.0, 0.0], [1.0, 1.0]	[0.0, 0.0], [1.0, 1.0]
c_2	[0.0, 0.0], [1.0, 1.0]	[0.0, 0.0], [1.0, 1.0]	[0.0, 0.0], [1.0, 1.0]	[0.0, 0.0], [1.0, 1.0]
v_1	[0.0, 0.0], [1.0, 1.0]	[0.0, 0.0], [1.0, 1.0]	[0.0, 0.0], [1.0, 1.0]	[0.0, 0.0], [1.0, 1.0]
v_2	[0.0, 0.0], [1.0, 1.0]	[0.0, 0.0], [1.0, 1.0]	[0.0, 0.0], [1.0, 1.0]	[0.0, 0.0], [1.0, 1.0]

An *absolute relation* I_U between them is given by I_U:

	(a_1, b_1)	(a_1, b_2)	(a_2, b_1)	(a_2, b_2)
h_1	[1.0, 1.0], [0.0, 0.0]	[1.0, 1.0], [0.0, 0.0]	[1.0, 1.0], [0.0, 0.0]	[1.0, 1.0], [0.0, 0.0]
h_2	[1.0, 1.0], [0.0, 0.0]	[1.0, 1.0], [0.0, 0.0]	[1.0, 1.0], [0.0, 0.0]	[1.0, 1.0], [0.0, 0.0]
h_3	[1.0, 1.0], [0.0, 0.0]	[1.0, 1.0], [0.0, 0.0]	[1.0, 1.0], [0.0, 0.0]	[1.0, 1.0], [0.0, 0.0]
c_1	[1.0, 1.0], [0.0, 0.0]	[1.0, 1.0], [0.0, 0.0]	[1.0, 1.0], [0.0, 0.0]	[1.0, 1.0], [0.0, 0.0]
c_2	[1.0, 1.0], [0.0, 0.0]	[1.0, 1.0], [0.0, 0.0]	[1.0, 1.0], [0.0, 0.0]	[1.0, 1.0], [0.0, 0.0]
v_1	[1.0, 1.0], [0.0, 0.0]	[1.0, 1.0], [0.0, 0.0]	[1.0, 1.0], [0.0, 0.0]	[1.0, 1.0], [0.0, 0.0]
v_2	[1.0, 1.0], [0.0, 0.0]	[1.0, 1.0], [0.0, 0.0]	[1.0, 1.0], [0.0, 0.0]	[1.0, 1.0], [0.0, 0.0]

Remark 6.34 For any $R \in \mathrm{MSSR}_U(A \times B)$, we have

  (i) $R \vee O_U = R$,
 (ii) $R \wedge O_U = O_U$,
(iii) $R \vee I_U = I_U$, and
 (iv) $R \wedge I_U = R$.

6.2 Various Types of Interval-Valued Intuitionistic Fuzzy Soft Multi-Set Relations

Definition 6.35 Let $\{U_\lambda : \lambda \in I\}$ be a collection of universes such that $\bigcap_{\lambda \in I} U_\lambda = \phi$ and let $\{E_{U_\lambda} : \lambda \in I\}$ be a collection of sets of parameters. Let $U = \prod_{\lambda \in I} \text{IVIFS}(U_\lambda)$ where $\text{IVIFS}(U_\lambda)$ denotes the set of all interval-valued intuitionistic fuzzy subsets of U_λ, $E = \prod_{\lambda \in I} E_{U_\lambda}$, and $A, B \subseteq E$. Let (F, A) and (G, B) be two interval-valued intuitionistic fuzzy soft multi-sets over U. Let $R \in \text{MSSR}_U$ $(A \times B)$ and $R = (H, A \times B)$. Then, R is called a *reflexive* IVIFSMS-relation if the relational matrix for R is a square matrix and for

$$(a_i, b_j) \in A \times B \text{ and } h_k^*$$

$\in U_\lambda$, we have, $\mu_{H(a_i, b_j)}(h_k^*)\big|_{(m,n)} = [1, 1]$ and $\gamma_{H(a_i, b_j)}(h_k^*)\big|_{(m,n)} = [0, 0]$

for $m = n = k$.

Example 6.36 Let us consider there are three universes U_1, U_2, and U_3.

Let $U_1 = \{h_1, h_2\}$, $U_2 = \{c_1\}$, and $U_3 = \{v_1\}$. Let $\{E_{U_1}, E_{U_2}, E_{U_3}\}$ be a collection of sets of decision parameters related to the above universes where

$$E_{U_1} = \{e_{U_1,1}, e_{U_1,2}, e_{U_1,3}\}, \quad E_{U_2} = \{e_{U_2,1}, e_{U_2,2}, e_{U_2,3}\},$$
$$E_{U_3} = \{e_{U_3,1}, e_{U_3,2}, e_{U_3,3}\}.$$

Let $U = \prod_{\lambda=1}^{3} \text{IVIFS}(U_\lambda)$, $E = \prod_{\lambda=1}^{3} E_{U_\lambda}$, and $A, B \subseteq E$, such that

$$A = \{a_1 = (e_{U_1,1}, e_{U_2,1}, e_{U_3,1}), a_2 = (e_{U_1,1}, e_{U_2,2}, e_{U_3,1})\},$$
$$B = \{b_1 = (e_{U_1,2}, e_{U_2,2}, e_{U_3,1}), b_2 = (e_{U_1,1}, e_{U_2,2}, e_{U_3,2})\}.$$

Then, a reflexive IVIFSMS-relation between them is

	(a_1, b_1)	(a_1, b_2)	(a_2, b_1)	(a_2, b_2)
h_1^*	[1.0, 1.0], [0.0, 0.0]	[0.6, 0.8], [0.1, 0.2]	[0.3, 0.5], [0.1, 0.3]	[0.5, 0.8], [0.1, 0.2]
h_2^*	[0.4, 0.6], [0.1, 0.4]	[1.0, 1.0], [0.0, 0.0]	[0.6, 0.8], [0.1, 0.2]	[0.3, 0.6], [0.3, 0.4]
h_3^*	[0.4, 0.6], [0.3, 0.4]	[0.2, 0.4], [0.2, 0.4]	[1.0, 1.0], [0.0, 0.0]	[0.2, 0.5], [0.4, 0.5]
h_4^*	[0.2, 0.5], [0.1, 0.2]	[0.1, 0.5], [0.3, 0.4]	[0.5, 0.7], [0.1, 0.2]	[1.0, 1.0], [0.0, 0.0]

Where $h_1^* = h_1$, $h_2^* = h_2$, $h_3^* = c_1$, and $h_4^* = v_1$

Definition 6.37 Let $\{U_\lambda : \lambda \in I\}$ be a collection of universes such that $\bigcap_{\lambda \in I} U_\lambda = \phi$ and let $\{E_{U_\lambda} : \lambda \in I\}$ be a collection of sets of parameters. Let $U = \prod_{\lambda \in I} \text{IVIFS}(U_\lambda)$ where $\text{IVIFS}(U_\lambda)$ denotes the set of all interval-valued intuitionistic fuzzy subsets of U_λ, $E = \prod_{\lambda \in I} E_{U_\lambda}$, and $A, B \subseteq E$. Let (F, A) and (G, B) be two interval-valued intuitionistic fuzzy soft multi-sets over U. Let $R \in \text{MSSR}_U$ $(A \times B)$ and $R = (H, A \times B)$. Then, R is called a *symmetric* IVIFSMS-relation if the relational matrix for R is a square matrix and if for each $(a_i, b_j) \in A \times B$ and $h_k^* \in U_\lambda$, $\exists (a_p, b_q) \in A \times B$ and $h_1^* \in U_\lambda$ such that

$$\mu_{H(a_i,b_j)}(h_k^*)\Big|_{(m,n)} = \mu_{H(a_p,b_q)}(h_1^*) \text{ and } \gamma_{H(a_i,b_j)}(h_k^*)\Big|_{(m,n)} = \gamma_{H(a_p,b_q)}(h_1^*)\Big|_{(n,m)}.$$

Example 6.38 Let us consider there are three universes U_1, U_2, and U_3.
Let $U_1 = \{h_1, h_2\}$, $U_2 = \{c_1\}$, and $U_3 = \{v_1\}$. Let $\{E_{U_1}, E_{U_2}, E_{U_3}\}$ be a collection of sets of decision parameters related to the above universes where

$$E_{U_1} = \{e_{U_1,1}, e_{U_1,2}, e_{U_1,3}\}, \quad E_{U_2} = \{e_{U_2,1}, e_{U_2,2}, e_{U_2,3}\},$$
$$E_{U_3} = \{e_{U_3,1}, e_{U_3,2}, e_{U_3,3}\}.$$

Let $U = \prod_{\lambda=1}^{3} \text{IVIFS}(U_\lambda)$, $E = \prod_{\lambda=1}^{3} E_{U_\lambda}$, and $A, B \subseteq E$, such that

$$A = \{a_1 = (e_{U_1,1}, e_{U_2,1}, e_{U_3,1}), a_2 = (e_{U_1,1}, e_{U_2,2}, e_{U_3,1})\},$$
$$B = \{b_1 = (e_{U_1,2}, e_{U_2,2}, e_{U_3,1}), b_2 = (e_{U_1,1}, e_{U_2,2}, e_{U_3,2})\}.$$

Then, a reflexive IVIFSMS-relation between them is

	(a_1, b_1)	(a_1, b_2)	(a_2, b_1)	(a_2, b_2)
h_1^*	[0.0, 0.2], [0.4, 0.6]	[0.6, 0.8], [0.1, 0.2]	[0.3, 0.5], [0.1, 0.3]	[0.2, 0.5], [0.1, 0.2]
h_2^*	[0.6, 0.8], [0.1, 0.2]	[0.3, 0.4], [0.5, 0.6]	[0.2, 0.4], [0.2, 0.4]	[0.3, 0.6], [0.3, 0.4]
h_3^*	[0.3, 0.5], [0.1, 0.3]	[0.2, 0.4], [0.2, 0.4]	[0.0, 0.0], [1.0, 1.0]	[0.2, 0.5], [0.4, 0.5]
h_4^*	[0.2, 0.5], [0.1, 0.2]	[0.3, 0.6], [0.3, 0.4]	[0.2, 0.5], [0.4, 0.5]	[0.4, 0.6], [0.2, 0.3]

Where $h_1^* = h_1$, $h_2^* = h_2$, $h_3^* = c_1$, and $h_4^* = v_1$

Definition 6.39 Let $\{U_\lambda : \lambda \in I\}$ be a collection of universes such that $\bigcap_{\lambda \in I} U_\lambda = \phi$ and let $\{E_{U_\lambda} : \lambda \in I\}$ be a collection of sets of parameters. Let $U = \prod_{\lambda \in I} \text{IVIFS}(U_\lambda)$ where $\text{IVIFS}(U_\lambda)$ denotes the set of all interval-valued intuitionistic fuzzy subsets of U_λ, $E = \prod_{\lambda \in I} E_{U_\lambda}$, and $A \subseteq E$. Let (F, A) and (G, A) be two interval-valued intuitionistic fuzzy soft multi-sets over U. Let R_1, $R_2 \in \text{MSSR}_U(A \times A)$ and $R_1 = (H, A \times A)$, $R_2 = (J, A \times A)$. Then, the composition of R_1 and R_2, denoted by $R_1 \circ R_2$, is defined by $R_1 \circ R_2 = (H \circ J, A \times A)$ where
$H \circ J: A \times A \to \text{IVIFS}(U)$ is defined as

$$(H \circ J)(a_i, a_j) = (\{\langle h_k^*, \mu_{(H \circ J)(a_i,a_j)}(h_k^*), \gamma_{(H \circ J)(a_i,a_j)}(h_k^*)\rangle : h_k^* \in U_\lambda\}\lambda \in I),$$

where

$$\mu_{(H \circ J)(a_i,a_j)}\left(h_k^*\right) = \left[\max_l \left(\min\left(\inf \mu_{H(a_i,a_l)}\left(h_k^*\right), \inf \mu_{J(a_l,a_j)}\left(h_k^*\right)\right)\right),\right.$$
$$\left.\max_l \left(\min\left(\sup \mu_{H(a_i,a_l)}\left(h_k^*\right), \sup \mu_{J(a_l,a_j)}\left(h_k^*\right)\right)\right)\right]$$

and

$$\gamma_{(H \circ J)(a_i,a_j)}\left(h_k^*\right) = \left[\min_l \left(\max\left(\inf \gamma_{H(a_i,a_l)}\left(h_k^*\right), \inf \gamma_{J(a_l,a_j)}\left(h_k^*\right)\right)\right),\right.$$
$$\left.\min_l, \left(\max\left(\sup \gamma_{H(a_i,a_l)}\left(h_k^*\right), \sup \gamma_{J(a_l,a_j)}\left(h_k^*\right)\right)\right)\right]$$
$$\text{for } (a_i, a_j) \in A \times A$$

Example 6.40 Let us consider there are three universes U_1, U_2, and U_3.

Let $U_1 = \{h_1, h_2\}$, $U_2 = \{c_1\}$, and $U_3 = \{v_1\}$. Let $\{E_{U_1}, E_{U_2}, E_{U_3}\}$ be a collection of sets of decision parameters related to the above universes where

$$E_{U_1} = \left\{e_{U_1,1}, e_{U_1,2}, e_{U_1,3}\right\}, \quad E_{U_2} = \left\{e_{U_2,1}, e_{U_2,2}, e_{U_2,3}\right\},$$
$$E_{U_3} = \left\{e_{U_3,1}, e_{U_3,2}, e_{U_3,3}\right\}.$$

Let $U = \prod_{\lambda=1}^{3} \text{IVIFS}(U_\lambda)$, $E = \prod_{\lambda=1}^{3} E_{U_\lambda}$, and $A \subseteq E$, such that

$$A = \{a_1 = (e_{U_1,1}, e_{U_2,1}, e_{U_3,1}), a_2 = (e_{U_1,1}, e_{U_2,2}, e_{U_3,1})\}.$$

Let $R_1, R_2 \in \text{MSSR}_U(A \times A)$ be defined by
R_1:

	(a_1, b_1)	(a_1, b_2)	(a_2, b_1)	(a_2, b_2)
h_1^*	([0.3, 0.4], [0.3, 0.4])	([0.2, 0.4], [0.3, 0.5])	([0.2, 0.5], [0.3, 0.4])	([0.2, 0.3], [0.3, 0.6])
h_2^*	([1.0, 1.0], [0.0, 0.0])	([0.1, 0.2], [0.0, 0.0])	([0.4, 0.5], [0.1, 0.3])	([0.4, 0.7], [0.1, 0.3])
h_3^*	([0.2, 0.6], [0.1, 0.4])	([0.2, 0.6], [0.1, 0.3])	([0.2, 0.3], [0.4, 0.6])	([0.2, 0.5], [0.2, 0.3])
h_4^*	([0.2, 0.4], [0.3, 0.5])	([0.3, 0.4], [0.4, 0.5])	([0.3, 0.4], [0.2, 0.3])	([0.0, 0.2], [0.4, 0.5])

R_2:

	(a_1, b_1)	(a_1, b_2)	(a_2, b_1)	(a_2, b_2)
h_1^*	([0.5, 0.8], [0.1, 0.2])	([0.2, 0.3], [0.3, 0.6])	([0.1, 0.4], [0.3, 0.5])	([0.2, 0.4], [0.2, 0.3])
h_2^*	([0.4, 0.5], [0.2, 0.4])	([0.4, 0.6], [0.2, 0.3])	([0.1, 0.5], [0.4, 0.5])	([0.4, 0.5], [0.1, 0.2])
h_3^*	([0.2, 0.3], [0.5, 0.6])	([0.3, 0.4], [0.4, 0.5])	([0.7, 0.8], [0.1, 0.2])	([0.3, 0.5], [0.3, 0.4])
h_4^*	([0.3, 0.5], [0.3, 0.4])	([0.3, 0.5], [0.2, 0.4])	([0.2, 0.4], [0.2, 0.3])	([0.3, 0.7], [0.1, 0.3])

Where $h_1^* = h_1$, $h_2^* = h_2$, $h_3^* = c_1$, and $h_4^* = v_1$.

Then, $R_1 o R_2$:

	(a_1, b_1)	(a_1, b_2)	(a_2, b_1)	(a_2, b_2)
h_1^*	([0.3, 0.4], [0.3, 0.4])	([0.2, 0.4], [0.3, 0.5])	([0.2, 0.5], [0.3, 0.4])	([0.2, 0.3], [0.3, 0.6])
h_2^*	([0.4, 0.5], [0.2, 0.4])	([0.1, 0.6], [0.1, 0.2])	([0.4, 0.5], [0.2, 0.4])	([0.4, 0.5], [0.1, 0.3])
h_3^*	([0.2, 0.6], [0.1, 0.3])	([0.2, 0.5], [0.3, 0.4])	([0.2, 0.5], [0.2, 0.3])	([0.2, 0.5], [0.3, 0.4])
h_4^*	([0.2, 0.4], [0.3, 0.5])	([0.3, 0.4], [0.3, 0.5])	([0.3, 0.4], [0.3, 0.4])	([0.3, 0.4], [0.2, 0.4])

Definition 6.41 Let $\{U_\lambda : \lambda \in I\}$ be a collection of universes such that $\bigcap_{\lambda \in I} U_\lambda = \phi$ and let $\{E_{U_\lambda} : \lambda \in I\}$ be a collection of sets of parameters. Let $U = \prod_{\lambda \in I} \text{IVIFS}(U_\lambda)$ where $\text{IVIFS}(U_\lambda)$ denotes the set of all interval-valued intuitionistic fuzzy subsets of U_λ, $E = \prod_{\lambda \in I} E_{U_\lambda}$ and $A \subseteq E$. Let (F, A) and (G, A) be two interval-valued intuitionistic fuzzy soft multi-sets over U. Let $R \in \text{MSSR}_U(A \times A)$. Then, R is called a *transitive* IVIFSMS-relation if $RoR \subseteq R$.

Example 6.42 Let us consider there are three universes U_1, U_2, and U_3.
Let $U_1 = \{h_1, h_2\}$, $U_2 = \{c_1\}$, and $U_3 = \{v_1\}$. Let $\{E_{U_1}, E_{U_2}, E_{U_3}\}$ be a collection of sets of decision parameters related to the above universes where

$$E_{U_1} = \{e_{U_1,1}, e_{U_1,2}, e_{U_1,3}\}, \quad E_{U_2} = \{e_{U_2,1}, e_{U_2,2}, e_{U_2,3}\},$$
$$E_{U_3} = \{e_{U_3,1}, e_{U_3,2}, e_{U_3,3}\}.$$

Let $U = \prod_{\lambda=1}^{3} \text{IVIFS}(U_\lambda)$, $E = \prod_{\lambda=1}^{3} E_{U_\lambda}$, and $A \subseteq E$, such that

$$A = \{a_1 = (e_{U_1,1}, e_{U_2,1}, e_{U_3,1}), a_2 = (e_{U_1,1}, e_{U_2,2}, e_{U_3,1})\}.$$

Let R:

	(a_1, b_1)	(a_1, b_2)	(a_2, b_1)	(a_2, b_2)
h_1^*	([0.3, 0.4], [0.3, 0.4])	([0.2, 0.4], [0.3, 0.6])	([0.2, 0.5], [0.3, 0.4])	([0.2, 0.4], [0.3, 0.6])
h_2^*	([1.0, 1.0], [0.0, 0.0])	([0.1, 0.2], [0.0, 0.0])	([0.4, 0.5], [0.1, 0.3])	([0.4, 0.7], [0.1, 0.3])
h_3^*	([0.2, 0.6], [0.1, 0.4])	([0.2, 0.6], [0.1, 0.3])	([0.2, 0.3], [0.4, 0.6])	([0.2, 0.5], [0.2, 0.3])
h_4^*	([0.3, 0.4], [0.3, 0.4])	([0.2, 0.4], [0.3, 0.5])	([0.2, 0.5], [0.3, 0.4])	([0.2, 0.4], [0.3, 0.5])

Then, RoR:

	(a_1, b_1)	(a_1, b_2)	(a_2, b_1)	(a_2, b_2)
h_1^*	([0.3, 0.4], [0.3, 0.4])	([0.2, 0.4], [0.3, 0.6])	([0.2, 0.4], [0.3, 0.4])	([0.2, 0.4], [0.3, 0.6])
h_2^*	([1.0, 1.0], [0.0, 0.0])	([0.1, 0.2], [0.0, 0.0])	([0.4, 0.5], [0.1, 0.3])	([0.4, 0.7], [0.1, 0.3])
h_3^*	([0.2, 0.6], [0.1, 0.4])	([0.2, 0.6], [0.1, 0.3])	([0.2, 0.3], [0.4, 0.6])	([0.2, 0.5], [0.2, 0.3])
h_4^*	([0.3, 0.4], [0.3, 0.4])	([0.2, 0.4], [0.3, 0.5])	([0.2, 0.5], [0.3, 0.4])	([0.2, 0.4], [0.3, 0.5])

Where $h_1^* = h_1$, $h_2^* = h_2$, $h_3^* = c_1$, and $h_4^* = v_1$

Then, clearly $RoR \subset R$. So R is a *transitive* IVIFSMS-relation.

6.3 Conclusion

In 1999, Molodtsov [9] introduced the concept of soft set theory as a general mathematical tool for dealing with uncertainties. Alkhazaleh et al. [2] in 2011 introduced the definition of soft multi-set as a generalisation of Molodtsov's soft set. In 2012, Alkhazaleh and Salleh [3] introduced the concept of fuzzy soft multi-set theory. In this chapter, we have introduced the concept of interval-valued intuitionistic fuzzy soft multi-sets and studied some of its properties and operations. Also we have defined interval-valued intuitionistic fuzzy soft multi-set relations. The basic properties of these relations are discussed. Also various types of these relations have been discussed in this chapter.

References

1. Ali, M.I., Feng, F., Liu, X., Min, W.K., Shabir, M.: On some new operations in soft set theory. Comput. Math. Appl. **57**(9), 1547–1553 (2009)
2. Alkhazaleh, S., Salleh, A.R., Hassan, N.: Soft multisets theory. Appl. Math. Sci. **5**, 3561–3573 (2011)
3. Alkhazaleh, S., Salleh, A.R.: Fuzzy soft multiset theory. Abstr. Appl. Anal. **2012**(350603) (2012)
4. Atanassov, K.: Intuitionistic fuzzy sets. Fuzzy Sets Syst. **20**, 87–96 (1986)
5. Atanassov, K., Gargov, G.: Interval-valued intuitionistic fuzzy sets. Fuzzy Sets Syst. **31**, 343–349 (1989)
6. Maji, P.K., Biswas, R., Roy, A.R.: Soft set theory. Comput. Math. Appl. **45**(4–5), 555–562 (2003)
7. Maji, P.K., Biswas, R., Roy, A.R.: Fuzzy soft sets. J. Fuzzy Math. **9**(3), 589–602 (2001)
8. Maji, P.K., Biswas, R., Roy, A.R.: Intuitionistic fuzzy soft sets. J. Fuzzy Math. **12**(3), 669–683 (2004)
9. Molodtsov, D.: Soft set theory-first results. Comput. Math. Appl. **37**(4–5), 19–31 (1999)
10. Zadeh, L.A.: Fuzzy sets. Inf. Control **8**, 338–353 (1965)

Chapter 7
Interval-Valued Neutrosophic Soft Sets

Abstract In this chapter, the concepts of interval-valued neutrosophic sets (IVNS in short), interval-valued neutrosophic soft sets (IVNSS in short) and IVNSS relations (IVNSS-relations in short) are proposed. The basic properties of IVNS-, IVNSS-, and IVNSS-relations are also presented and discussed. Also various types of IVNSS-relations are presented. Finally, a solution to a decision-making problem using IVNSS-relation is presented.

Keywords Soft set · Neutrosophic set · Neutrosophic soft set · IVNS · IVNSS · IVNSS-relations

There are many complicated problems in economics, engineering, environmental science, and social science which cannot be solved by the well-known methods of classical mathematics (as various types of uncertainties are presented in these problems). To handle situations like these, many tools have been suggested. Some of them are probability theory, fuzzy set theory [9], rough set theory [6], etc. The traditional fuzzy set is characterised by the membership value or the grade of membership value. Sometimes it may be very difficult to assign the membership value for fuzzy sets. Interval-valued fuzzy sets [8] were proposed as a natural extension of fuzzy sets and were proposed independently by Zadeh [10] to capture the uncertainty of grade of membership value. In some real-life problems in expert system, belief system, information fusion and so on, we must consider the truth-membership as well as the falsity-membership for proper description of an object in uncertain, ambiguous environment. Neither the fuzzy sets nor the interval-valued fuzzy set is appropriate for such a situation. Intuitionistic fuzzy sets introduced by Atanassov [2] are appropriate for such a situation. The intuitionistic fuzzy sets can only handle the incomplete information considering both the truth-membership (or simply membership) and the falsity-membership (or non-membership) values. It does not handle the indeterminate and inconsistent information which exists in belief system. In 1999, Molodstov [5] introduced soft set theory which is completely a new approach for modelling vagueness and uncertainties. Research works on soft set theory are progressing rapidly. Maji et al. [3] defined several operations on soft set theory. Based on the analysis of several operations on soft sets introduced in [3], Ali et al. [1] presented some new algebraic operations for soft sets and proved that certain De Morgan's law holds in soft set theory with respect to these new definitions. Smarandache [7] introduced the concept of neutrosophic set which is a mathematical tool for handling problems involving imprecise, indeterminacy, and inconsistent data. Maji [4] introduced the concept of neutrosophic soft set and established some operations on these sets. In this chapter we introduce the concept of interval-valued neutrosophic sets (IVNS in short), interval-valued neutrosophic soft sets (IVNSS in short) and IVNSS relations (IVNSS-relations in short). The basic properties of IVNSS- and IVNSS-relations are also presented and discussed. Also various types of IVNSS-relations are presented.

A neutrosophic set A on the universe of discourse U is defined as $A = \{\langle x, \mu_A(x), \gamma_A(x), \delta_A(x) \rangle : x \in U\}$, where $\mu_A, \gamma_A, \delta_A : U \rightarrow]^-0, 1^+[$ are the functions such that the condition: $\forall x \in U, {}^-0 \leq \mu_A(x) + \gamma_A(x) + \delta_A(x) \leq 3^+$ is satisfied.

Here, $\mu_A(x), \gamma_A(x), \delta_A(x)$ represent the truth-membership, indeterminacy-membership, and falsity-membership, respectively, of the element $x \in U$. Informally, an infinitesimal is an infinitely small number. Formally, n is said to be infinitesimal if and only if for all positive integers n one has $|n| < f$. Let $\varepsilon > 0$ be such infinitesimal number. The hyper-real number is an extension of the real number set, which includes classes of infinite numbers and classes of infinitesimal numbers. We consider the non-standard finite numbers $1^+ = 1 + \varepsilon$ where 1 is its standard part and ε is its non-standard part and $0^- = 0 - \varepsilon$, where 0 is its standard part and ε is non-standard part. We call $]^-0, 1^+[$ a non-standard unit interval.

Let T, I, F be standard or non-standard real subsets of $]^-0, 1^+[$ with $\sup T = t_{\sup}$, $\inf T = t_{\inf}$, $\sup I = i_{\sup}$, $\inf I = i_{\inf}$, $\sup F = f_{\sup}$, $\inf F = f_{\inf}$, and $n_{\sup} = t_{\sup} + i_{\sup} + f_{\sup}$, $n_{\inf} = t_{\inf} + i_{\inf} + f_{\inf}$. The sets T, I, F are not necessarily intervals, but may be any real sub-unitary subsets: discrete or continuous, single element, finite or (countably or uncountably) infinite, union or intersection of various subsets, etc. They may also overlap. The real subsets could represent the relative errors in determining t, i, f (in the case where the subsets T, I, F are reduced to points). T, I, F are the neutrosophic components represent the truth value, indeterminacy value, and falsehood value, respectively, referring to neutrosophy, neutrosophic logic, neutrosophic set, neutrosophic probability, and neutrosophic statistics.

A logic in which each proposition is estimated to have the percentage of truth in a subset T, the percentage of indeterminacy in a subset I, and the percentage of falsity in a subset F where T, I, F are defined above is called neutrosophic logic. T, I, F are standard or non-standard subsets of the non-standard interval $]^-0, 1^+[$, where $n_{\inf} = \inf T + \inf I + \inf F \geq {}^-0$ and $n_{\sup} = \sup T + \sup I + \sup F \leq 3^+$.

From philosophical point of view, the neutrosophic set takes the value from real standard or non-standard subsets of $]^-0, 1^+[$. But in real-life application in scientific and engineering problems, it is difficult to use neutrosophic set with value from real standard or non-standard subset of $]^-0, 1^+[$. Hence, we consider the neutrosophic set which takes the value from the subset of $[0, 1]$.

Pei and Miao showed $A \times B$ as basic binary operation, where $A, B \subseteq E$ (where E is the parameter set). Ali et al. also defined some new operations. Ali introduced soft binary relation on a set X.

Definition 7.1 The union of two soft sets (F, A) and (G, B) over the common universe U is the soft set (H, C), where $C = A \cup B$ and $\forall e \in C$

$$H(e) = \begin{cases} F(e) & \text{if } e \in A - B \\ G(e) & \text{if } e \in B - A \\ F(e) \cup G(e) & \text{if } e \in A \cap B \end{cases}$$

We write $(F, A) \widetilde{\cup} (G, B) = (H, C)$.

Definition 7.2 The intersection of two soft sets (F, A) and (G, B) over the common universe U is the soft set (H, C), where $C = A \cup B$ and $\forall e \in C$

$$H(e) = \begin{cases} F(e) & \text{if } e \in A - B \\ G(e) & \text{if } e \in B - A \\ F(e) \cap G(e) & \text{if } e \in A \cap B \end{cases}$$

We write $(F, A) \widetilde{\cap} (G, B) = (H, C)$.

Definition 7.3 The complement of a soft set (F, A) is denoted by $(F, A)^c$ and is defined by $(F, A)^c = (F^c, \rceil A)$, where $F^c: \rceil A \rightarrow P(U)$ is a mapping given by $F^c(\alpha) = $ complement of $F(\alpha) = U - F(\alpha)$ for $\alpha \in \rceil A$.

Definition 7.4 A neutrosophic set A on the universe of discourse U is defined as $A = \{\langle x, \mu_A(x), \gamma_A(x), \delta_A(x) \rangle : x \in U\}$, where $\mu_A, \gamma_A, \delta_A : U \rightarrow]^-0, 1^+[$ are the functions such that the condition: $\forall x \in U, \ ^-0 \leq \mu_A(x) + \gamma_A(x) + \delta_A(x) \leq 3^+$ is satisfied.

Smarandache [7] applied neutrosophic sets in many directions after giving examples of neutrosophic sets. Then, he introduced the neutrosophic set operations, namely complement, union, intersection, difference, and Cartesian product in Smarandache [7].

Definition 7.5 Let U be an initial universe, E be a set of parameters, and $A \subseteq E$. Let $NP(U)$ denotes the set of all neutrosophic sets of U. Then, the pair (F, A) is termed to be the *neutrosophic soft set* over U, where F is a mapping given by $F: A \rightarrow NP(U)$.

7.1 Interval-Valued Neutrosophic Sets

In this section, we introduce the concept of IVNS and study their basic properties.

Definition 7.6 An IVNS A on the universe of discourse U is defined as: $A = \{\langle x, \mu_A(x), \gamma_A(x), \delta_A(x)\rangle : x \in U\}$, where $\mu_A, \gamma_A, \delta_A : U \rightarrow]^-0, 1^+[$ are functions such that the condition: $\forall x \in U,\ ^-0 \leq \sup \mu_A(x) + \sup \gamma_A(x) + \sup \delta_A(x) \leq 3^+$ is satisfied.

In real-life applications it is difficult to use IVNS with interval-value from real standard or non-standard subset of Int($]^-0, 1^+[$). Hence we consider the IVNS that takes the interval-value from the subset of Int($[0, 1]$). The set of all IVNS on U is denoted by IVNS^U.

Definition 7.7 Let A and B be two IVNS on U defined by $A = \{\langle x, \mu_A(x), \gamma_A(x), \delta_A(x)\rangle : x \in U\}$ and $B = \{\langle x, \mu_B(x), \gamma_B(x), \delta_B(x)\rangle : x \in U\}$. Then

1. A is called a subset of B, denoted by $A \subseteq B$, if

$$\mu_A(x) \leq \mu_B(x),\ \gamma_A(x) \leq \gamma_B(x),\ \delta_A(x) \leq \delta_B(x) \quad \text{for } x \in U.$$

2. Their *union* is denoted by $A \cup B$ and is defined by the IVNS

$$A \cup B = \{\langle x, \mu_A(x) \vee \mu_B(x), \gamma_A(x) \square \gamma_B(x), \delta_A(x) \wedge \delta_B(x)\rangle : x \in U\}$$

where for $x \in U$

$$\mu_A(x) \vee \mu_B(x) = [\max(\inf \mu_A(x), \inf \mu_B(x)), \max(\sup \mu_A(x), \sup \mu_B(x))]$$

$$\gamma_A(x) \square \gamma_B(x) = \left[\frac{\inf \gamma_A(x) + \inf \gamma_B(x)}{2}, \frac{\sup \gamma_A(x) + \sup \gamma_B(x)}{2}\right]$$

$$\delta_A(x) \wedge \delta_B(x) = [\min(\inf \delta_A(x), \inf \delta_B(x)), \min(\sup \delta_A(x), \sup \delta_B(x))].$$

3. Their *intersection* is denoted by $A \cap B$ and is defined by the IVNS $A \cap B = \{\langle x, \mu_A(x) \wedge \mu_B(x), \gamma_A(x) \square \gamma_B(x), \delta_A(x) \vee \delta_B(x)\rangle : x \in U\}$ where for $x \in U$

$$\mu_A(x) \wedge \mu_B(x) = [\min(\inf \mu_A(x), \inf \mu_B(x)), \min(\sup \mu_A(x), \sup \mu_B(x))]$$

$$\gamma_A(x) \square \gamma_B(x) = \left[\frac{\inf \gamma_A(x) + \inf \gamma_B(x)}{2}, \frac{\sup \gamma_A(x) + \sup \gamma_B(x)}{2}\right]$$

$$\delta_A(x) \vee \delta_B(x) = [\max(\inf \delta_A(x), \inf \delta_B(x)), \max(\sup \delta_A(x), \sup \delta_B(x))].$$

4. The *complement* of A is denoted by A^c and is defined by the IVNS $A^c = \{\langle x, \delta_A(x), \gamma_A(x), \mu_A(x)\rangle : x \in U\}$.

Theorem 7.8 *Let* $A, B, C \in \text{IVNS}^U$. *Then*

1. $A \cup A = A$
2. $A \cap A = A$
3. $A \cup B = B \cup A$
4. $A \cap B = B \cap A$
5. $(A \cup B)^c = A^c \cap B^c$
6. $(A \cap B)^c = A^c \cup B^c$
7. $(A \cup B) \cup C = A \cup (B \cup C)$
8. $(A \cap B) \cap C = A \cap (B \cap C)$
9. $A \cup (B \cap C) = (A \cup B) \cap (A \cup C)$
10. $A \cap (B \cup C) = (A \cap B) \cup (A \cap C)$

Proof Straight forward. $\square$

7.2 Interval-Valued Neutrosophic Soft Sets

In this section, we introduce the concept of IVNSS and study their basic properties.

Definition 7.9 Let U be an universe set, E be a set of parameters and $A \subseteq E$. Let IVNS^U denote the set of all IVNS of U. Then, the pair (f, A) is called an IVNSS over U, where f is a mapping given by $f: A \rightarrow \text{IVNS}^U$.

The collection of all IVNSS over U is denoted by IVNSS^U.

Example 7.10 Let $U = \{h_1, h_2, h_3, h_4, h_5\}$ be the set of five houses and $A = \{e_1(\text{expensive}), e_2(\text{wooden}), e_3(\text{beautiful}), e_4(\text{in the green surroundings})\}$.
    Then, the tabular representation of an IVNSS (f, A) can be given by:

U	e_1	e_2	e_3	e_4
h_1	([0.2, 0.4], [0.3, 0.5], [0.6, 0.8])	([0.2, 0.3], [0.5, 0.6], [0.7, 0.8])	([0.5, 0.8], [0.1, 0.2], [0.3, 0.4])	([0.1, 0.3], [0.4, 0.6], [0.1, 0.2])
h_2	([0.5, 0.7], [0.2, 0.4], [0.3, 0.6])	([0.7, 0.9], [0.2, 0.3], [0.1, 0.2])	([0.2, 0.3], [0.6, 0.8], [0.5, 0.7])	([0.3, 0.6], [0.2, 0.4], [0.5, 0.9])
h_3	([0.4, 0.6], [0.1, 0.3], [0.4, 0.5])	([0.5, 0.7], [0.4, 0.6], [0.2, 0.3])	([0.4, 0.6], [0.3, 0.5], [0.7, 0.8])	([0.4, 0.7], [0.1, 0.3], [0.3, 0.6])
h_4	([0.6, 0.8], [0.4, 0.6], [0.1, 0.2])	([0.4, 0.5], [0.1, 0.3], [0.4, 0.7])	([0.4, 0.7], [0.2, 0.3], [0.4, 0.5])	([0.2, 0.3], [0.5, 0.6], [0.7, 0.8])
h_5	([0.5, 0.9], [0.5, 0.6], [0.2, 0.4])	([0.2, 0.3], [0.4, 0.6], [0.2, 0.5])	([0.1, 0.4], [0.5, 0.6], [0.1, 0.4])	([0.5, 0.8], [0.1, 0.2], [0.3, 0.5])

Definition 7.11 Let U be an universe set and E be a set of parameters. Let $(f, A), (g, B) \in \mathrm{IVNSS}^U$, where $f : A \to \mathrm{IVNS}^U$ is defined by $f(a) = \{ \langle x, \mu_{f(a)}(x),$ $\gamma_{f(a)}(x), \delta_{f(a)}(x) \rangle : x \in U \}$ and $g : B \to \mathrm{IVNS}^U$ is defined by $g(b) = \{ \langle x, \mu_{g(b)}(x),$ $\gamma_{g(b)}(x), \delta_{g(b)}(x) \rangle : x \in U \}$, where $\mu_{f(a)}(x), \gamma_{f(a)}(x), \delta_{f(a)}(x), \mu_{g(b)}(x), \gamma_{g(b)}(x), \delta_{g(b)}$ $(x) \in \mathrm{Int}([0, 1])$ for $x \in U$.

Then,

(i) Their *union*, denoted by $(f, A) \cup (g, B) = (h, C)$ (say), is an IVNSS over U, where $C = A \cup B$ and for $e \in C$, $h : C \to \mathrm{IVNS}^U$ is defined by

$$h(e) = \left\{ \left\langle x, \mu_{h(e)}(x), \gamma_{h(e)}(x), \delta_{h(e)}(x) \right\rangle : x \in U \right\}$$

where for $x \in U$,

$$\mu_{h(e)}(x) = \begin{cases} \mu_{f(e)}(x) & \text{if } e \in A - B \\ \mu_{g(e)}(x) & \text{if } e \in B - A \\ \mu_{f(e)}(x) \vee \mu_{g(e)}(x) & \text{if } e \in A \cap B \end{cases}$$

$$\gamma_{h(e)}(x) = \begin{cases} \gamma_{f(e)}(x) & \text{if } e \in A - B \\ \gamma_{g(e)}(x) & \text{if } e \in B - A \\ \gamma_{f(e)} \square \gamma_{g(e)} & \text{if } e \in A \cap B \end{cases}$$

$$\delta_{h(e)}(x) = \begin{cases} \delta_{f(e)}(x) & \text{if } e \in A - B \\ \delta_{g(e)}(x) & \text{if } e \in B - A \\ \delta_{f(e)}(x) \wedge \delta_{g(e)}(x) & \text{if } e \in A \cap B \end{cases}$$

(ii) Their *intersection*, denoted by $(f, A) \cap (g, B) = (h, C)$ (say), is an IVNSS of over U, where $C = A \cup B$ and for $e \in C$, $h : C \to \mathrm{IVNS}^U$ is defined by $h(e) = \left\{ \left\langle x, \mu_{h(e)}(x), \gamma_{h(e)}(x), \delta_{h(e)}(x) \right\rangle : x \in U \right\}$, where for $x \in U$,

$$\mu_{h(e)}(x) = \begin{cases} \mu_{f(e)}(x) & \text{if } e \in A - B \\ \mu_{g(e)}(x) & \text{if } e \in B - A \\ \mu_{f(e)}(x) \wedge \mu_{g(e)}(x) & \text{if } e \in A \cap B \end{cases}$$

$$\gamma_{h(e)}(x) = \begin{cases} \gamma_{f(e)}(x) & \text{if } e \in A - B \\ \gamma_{g(e)}(x) & \text{if } e \in B - A \\ \gamma_{f(e)} \square \gamma_{g(e)} & \text{if } e \in A \cap B \end{cases}$$

$$\delta_{h(e)}(x) = \begin{cases} \delta_{f(e)}(x) & \text{if } e \in A - B \\ \delta_{g(e)}(x) & \text{if } e \in B - A \\ \delta_{f(e)}(x) \vee \delta_{g(e)}(x) & \text{if } e \in A \cap B \end{cases}$$

(iii) The *complement of* (f, A), denoted by $(f, A)^c$, is an IVNSS over U and is defined as $(f, A)^c = (f^c, A)$, where $f^c{:}A \rightarrow \text{IVNS}^U$ is defined by $f^c(a) = \left\{ \langle x, \delta_{f(a)}(x), \gamma_{f(a)}(x), \mu_{f(a)}(x) \rangle {:} x \in U \right\}$, where for $a \in A$.

Example 7.12 Let $U = \{h_1, h_2, h_3, h_4, h_5\}$ be the set of five houses and $A = \{e_1(\text{expensive}), e_2(\text{wooden}), e_3(\text{beautiful})\}$.

Then, the tabular representation of an IVNSS (F, A) can be given by:

U	e_1	e_2	e_3
h_1	([0.2, 0.4], [0.3, 0.5], [0.6, 0.8])	([0.2, 0.3], [0.5, 0.6], [0.7, 0.8])	([0.5, 0.8], [0.1, 0.2], [0.3, 0.4])
h_2	([0.5, 0.7], [0.2, 0.4], [0.3, 0.6])	([0.7, 0.9], [0.2, 0.3], [0.1, 0.2])	([0.2, 0.3], [0.6, 0.8], [0.5, 0.7])
h_3	([0.4, 0.6], [0.1, 0.3], [0.4, 0.5])	([0.5, 0.7], [0.4, 0.6], [0.2, 0.3])	([0.4, 0.6], [0.3, 0.5], [0.7, 0.8])
h_4	([0.6, 0.8], [0.4, 0.6], [0.1, 0.2])	([0.4, 0.5], [0.1, 0.3], [0.4, 0.7])	([0.4, 0.7], [0.2, 0.3], [0.4, 0.5])
h_5	([0.5, 0.9], [0.5, 0.6], [0.2, 0.4])	([0.2, 0.3], [0.4, 0.6], [0.2, 0.5])	([0.1, 0.4], [0.5, 0.6], [0.1, 0.4])

Let $B = \{e_1(\text{expensive}), e_4(\text{in green surroundings})\}$.
Then the tabular representation of an IVNSS (G, B) can be given by:

U	e_1	e_4
h_1	([0.3, 0.7], [0.4, 0.8], [0.6, 0.9])	([0.1, 0.3], [0.4, 0.6], [0.1, 0.2])
h_2	([0.2, 0.4], [0.5, 0.7], [0.6, 0.8])	([0.3, 0.6], [0.2, 0.4], [0.5, 0.9])
h_3	([0.7, 0.8], [0.2, 0.3], [0.5, 0.7])	([0.4, 0.7], [0.1, 0.3], [0.3, 0.6])
h_4	([0.2, 0.9], [0.3, 0.6], [0.5, 0.6])	([0.2, 0.3], [0.5, 0.6], [0.7, 0.8])
h_5	([0.2, 0.8], [0.7, 0.9], [0.6, 0.8])	([0.5, 0.8], [0.1, 0.2], [0.3, 0.5])

Let $(f, A) \cup (G, B) = (h, C)$ then $C = \{e_1, e_2, e_3, e_4\}$. The tabular representation of the IVNSS (h, C) is given as follows:

U	e_1	e_2	e_3	e_4
h_1	([0.3, 0.7], [0.35, 0.65], [0.6, 0.8])	([0.2, 0.3], [0.5, 0.6], [0.7, 0.8])	([0.5, 0.8], [0.1, 0.2], [0.3, 0.4])	([0.1, 0.3], [0.4, 0.6], [0.1, 0.2])
h_2	([0.5, 0.7], [0.35, 0.55], [0.3, 0.6])	([0.7, 0.9], [0.2, 0.3], [0.1, 0.2])	([0.2, 0.3], [0.6, 0.8], [0.5, 0.7])	([0.3, 0.6], [0.2, 0.4], [0.5, 0.9])
h_3	([0.7, 0.8], [0.15, 0.3], [0.4, 0.5])	([0.5, 0.7], [0.4, 0.6], [0.2, 0.3])	([0.4, 0.6], [0.3, 0.5], [0.7, 0.8])	([0.4, 0.7], [0.1, 0.3], [0.3, 0.6])
h_4	([0.6, 0.9], [0.35, 0.6], [0.1, 0.2])	([0.4, 0.5], [0.1, 0.3], [0.4, 0.7])	([0.4, 0.7], [0.2, 0.3], [0.4, 0.5])	([0.2, 0.3], [0.5, 0.6], [0.7, 0.8])
h_5	([0.5, 0.9], [0.6, 0.75], [0.2, 0.4])	([0.2, 0.3], [0.4, 0.6], [0.2, 0.5])	([0.1, 0.4], [0.5, 0.6], [0.1, 0.4])	([0.5, 0.8], [0.1, 0.2], [0.3, 0.5])

Let $(f,A) \cap (G,B) = (p,D)$ then $D = \{e_1, e_2, e_3, e_4\}$. The tabular representation of the IVNSS (p,D) is given as follows:

U	e_1	e_2	e_3	e_4
h_1	([0.2, 0.4], [0.35, 0.65], [0.6, 0.9])	([0.2, 0.3], [0.5, 0.6], [0.7, 0.8])	([0.5, 0.8], [0.1, 0.2], [0.3, 0.4])	([0.1, 0.3], [0.4, 0.6], [0.1, 0.2])
h_2	([0.2, 0.4], [0.35, 0.55], [0.6, 0.8])	([0.7, 0.9], [0.2, 0.3], [0.1, 0.2])	([0.2, 0.3], [0.6, 0.8], [0.5, 0.7])	([0.3, 0.6], [0.2, 0.4], [0.5, 0.9])
h_3	([0.4, 0.6], [0.15, 0.3], [0.5, 0.7])	([0.5, 0.7], [0.4, 0.6], [0.2, 0.3])	([0.4, 0.6], [0.3, 0.5], [0.7, 0.8])	([0.4, 0.7], [0.1, 0.3], [0.3, 0.6])
h_4	([0.2, 0.8], [0.35, 0.6], [0.5, 0.6])	([0.4, 0.5], [0.1, 0.3], [0.4, 0.7])	([0.4, 0.7], [0.2, 0.3], [0.4, 0.5])	([0.2, 0.3], [0.5, 0.6], [0.7, 0.8])
h_5	([0.2, 0.8], [0.6, 0.75], [0.6, 0.8])	([0.2, 0.3], [0.4, 0.6], [0.2, 0.5])	([0.1, 0.4], [0.5, 0.6], [0.1, 0.4])	([0.5, 0.8], [0.1, 0.2], [0.3, 0.5])

We have $(f,A)^c = (f^c, A)$ then. The tabular representation of the IVNSS $(f,A)^c$ is given as:

U	e_1	e_2	e_3
h_1	([0.6, 0.8], [0.3, 0.5], [0.2, 0.4])	([0.7, 0.8], [0.5, 0.6], [0.2, 0.3])	([0.3, 0.4], [0.1, 0.2], [0.5, 0.8])
h_2	([0.3, 0.6], [0.2, 0.4], [0.5, 0.7])	([0.1, 0.2], [0.2, 0.3], [0.7, 0.9])	([0.5, 0.7], [0.6, 0.8], [0.2, 0.3])
h_3	([0.4, 0.5], [0.1, 0.3], [0.4, 0.6])	([0.2, 0.3], [0.4, 0.6], [0.5, 0.7])	([0.7, 0.8], [0.3, 0.5], [0.4, 0.6])
h_4	([0.1, 0.2], [0.4, 0.6], [0.6, 0.8])	([0.4, 0.7], [0.1, 0.3], [0.4, 0.5])	([0.4, 0.5], [0.2, 0.3], [0.4, 0.7])
h_5	([0.2, 0.4], [0.5, 0.6], [0.5, 0.9])	([0.2, 0.5], [0.4, 0.6], [0.2, 0.3])	([0.1, 0.4], [0.5, 0.6], [0.1, 0.4])

Theorem 7.13 *Let U be an initial universe and $(f,A),(g,B),(h,C) \in \text{IVNSS}^U$ then*

1. $(f,A) \cup (f,A) = (f,A)$
2. $(f,A) \cap (f,A) = (f,A)$
3. $((f,A) \cup (g,A))^c = (f,A)^c \cap (g,A)^c$ *for $B = A$*
4. $((f,A) \cap (g,A))^c = (f,A)^c \cup (g,A)^c$ *for $B = A$*
5. $((f,A) \cup (g,B)) \cup (h,C) = (f,A) \cup ((g,B) \cup (h,C))$
6. $((f,A) \cap (g,B)) \cap (h,C) = (f,A) \cap ((g,B) \cap (h,C))$
7. $(f,A) \cup ((g,B) \cap (h,C)) = ((f,A) \cup (g,B)) \cap ((f,A) \cup (h,C))$
8. $(f,A) \cap ((g,B) \cup (h,C)) = ((f,A) \cap (g,B)) \cup ((f,A) \cap (h,C))$

Proof Let $f: A \to \text{IVNS}^U$, $g: B \to \text{IVNS}^U$, $h: C \to \text{IVNS}^U$ be defined by

$$f(a) = \left\{ \left\langle x, \mu_{f(a)}(x), \gamma_{f(a)}(x), \delta_{f(a)}(x) \right\rangle : x \in U \right\} \quad \text{for } a \in A$$

$$g(b) = \left\{ \left\langle x, \mu_{g(b)}(x), \gamma_{g(b)}(x), \delta_{g(b)}(x) \right\rangle : x \in U \right\} \quad \text{for } b \in B$$

$$h(c) = \left\{ \left\langle x, \mu_{h(c)}(x), \gamma_{h(c)}(x), \delta_{h(c)}(x) \right\rangle : x \in U \right\} \quad \text{for } c \in C, \text{ respectively.}$$

(1)–(2) Straight forward.

(3) Let $(f, A) \cup (g, A) = (s, A)$, where $s: A \to \text{IVNS}^U$ is given by for $e \in A$

$$s(e) = \left\{ \left\langle x, \mu_{s(e)}(x), \gamma_{s(e)}(x), \delta_{s(e)}(x) \right\rangle : x \in U \right\}$$

where

$$\mu_{s(e)}(x) = \mu_{f(e)}(x) \vee \mu_{g(e)}(x)$$
$$\gamma_{s(e)}(x) = \gamma_{f(e)}(x) \square \gamma_{g(e)}(x)$$
$$\delta_{s(e)}(x) = \delta_{f(e)}(x) \wedge \delta_{g(e)}(x)$$

Now $((f, A) \cup (g, A))^c = (s, A)^c = (s^c, A)$, where $s^c(e) = \{\langle x, \delta_{s(e)}(x), \gamma_{s(e)}(x), \mu_{s(e)}(x) \rangle : x \in U\}$.

Now $(f, A)^c = (f^c, A)$, where $f^c(e) = \{\langle x, \delta_{f(e)}(x), \gamma_{f(e)}(x), \mu_{f(e)}(x) \rangle : x \in U\}$ for $e \in A$ and $(g, A)^c = (g^c, A)$, where $g^c(e) = \left\{ \left\langle x, \delta_{g(e)}(x), \gamma_{g(e)}(x), \mu_{g(e)}(x) \right\rangle : x \in U \right\}$ for $e \in A$ therefore $(f, A)^c \cap (g, A)^c = (t, A)$ (Say) where $t(e) = \{\langle x, \delta_{f(e)}(x) \wedge \delta_{g(e)}(x), \gamma_{f(e)}(x) \square \gamma_{g(e)}(x), \mu_{f(e)}(x) \vee \mu_{g(e)}(x) \rangle : x \in U\}$ for $e \in A$

Consequently, $((f, A) \cup (g, A))^c = (f, A)^c \cap (g, A)^c$.

(4) Similar to (3).

(5) Let $(f, A) \cup (g, B) = (s, D)$ and $D = A \cup B$. Then for $x \in U$ and $d \in D$

$$\mu_{s(d)}(x) = \begin{cases} \mu_{f(d)}(x) & \text{if } d \in A - B \\ \mu_{g(d)}(x) & \text{if } d \in B - A \\ \mu_{f(d)}(x) \vee \mu_{g(d)}(x) & \text{if } d \in A \cap B \end{cases}$$

$$\gamma_{s(d)}(x) = \begin{cases} \gamma_{f(d)}(x) & \text{if } d \in A - B \\ \gamma_{g(d)}(x) & \text{if } d \in B - A \\ \gamma_{f(e)}(x) \square \gamma_{g(e)}(x) & \text{if } d \in A \cap B \end{cases}$$

$$\delta_{s(d)}(x) = \begin{cases} \delta_{f(d)}(x) & \text{if } d \in A - B \\ \delta_{g(d)}(x) & \text{if } d \in B - A \\ \delta_{f(d)}(x) \wedge \delta_{g(d)}(x) & \text{if } d \in A \cap B \end{cases}$$

Now   let $((f, A) \cup (g, B)) \cup (h, C) = (s, D) \cup (h, C) = (t, E)$ where $E = D \cup C = A \cup B \cup C$. Then for $x \in U$ and $e \in E$ we have

$$\mu_{t(e)}(x) = \begin{cases} \mu_{f(e)}(x) & \text{if } e \in A - B - C \\ \mu_{g(e)}(x) & \text{if } e \in B - C - A \\ \mu_{h(e)}(x) & \text{if } e \in C - A - B \\ \mu_{f(e)}(x) \vee \mu_{g(e)}(x) & \text{if } e \in (A \cap B) - C \\ \mu_{g(e)}(x) \vee \mu_{h(e)}(x) & \text{if } e \in (B \cap C) - A \\ \mu_{h(e)}(x) \vee \mu_{f(e)}(x) & \text{if } e \in (C \cap A) - B \\ \mu_{f(e)}(x) \vee \mu_{g(e)}(x) \vee \mu_{h(e)}(x) & \text{if } e \in A \cap B \cap C \end{cases}$$

$$\gamma_{t(e)}(x) = \begin{cases} \gamma_{f(e)}(x) & \text{if } e \in A - B - C \\ \gamma_{g(e)}(x) & \text{if } e \in B - C - A \\ \gamma_{h(e)}(x) & \text{if } e \in C - A - B \\ \gamma_{f(e)}(x) \square \gamma_{g(e)}(x) & \text{if } e \in (A \cap B) - C \\ \gamma_{g(e)}(x) \square \gamma_{h(e)}(x) & \text{if } e \in (B \cap C) - A \\ \gamma_{h(e)}(x) \square \gamma_{f(e)}(x) & \text{if } e \in (C \cap A) - B \\ \gamma_{f(e)}(x) \square \gamma_{g(e)}(x) \square \gamma_{h(e)}(x) & \text{if } e \in A \cap B \cap C \end{cases}$$

$$\delta_{t(e)}(x) = \begin{cases} \delta_{f(e)}(x) & \text{if } e \in A - B - C \\ \delta_{g(e)}(x) & \text{if } e \in B - C - A \\ \delta_{h(e)}(x) & \text{if } e \in C - A - B \\ \delta_{f(e)}(x) \wedge \delta_{g(e)}(x) & \text{if } e \in (A \cap B) - C \\ \delta_{g(e)}(x) \wedge \delta_{h(e)}(x) & \text{if } e \in (B \cap C) - A \\ \delta_{h(e)}(x) \wedge \delta_{f(e)}(x) & \text{if } e \in (C \cap A) - B \\ \delta_{f(e)}(x) \wedge \delta_{g(e)}(x) \wedge \delta_{h(e)}(x) & \text{if } e \in A \cap B \cap C \end{cases}$$

Again, let $(g, B) \cup (h, C) = (u, J)$, where $J = B \cup C$. Then for $x \in U$ and $j \in J$, we have

$$\mu_{u(j)}(x) = \begin{cases} \mu_{g(j)}(x) & \text{if } j \in B - C \\ \mu_{h(j)}(x) & \text{if } j \in C - B \\ \mu_{g(j)}(x) \vee \mu_{h(j)}(x) & \text{if } j \in B \cap C \end{cases}$$

$$\gamma_{u(j)}(x) = \begin{cases} \gamma_{g(j)}(x) & \text{if } j \in B - C \\ \gamma_{h(j)}(x) & \text{if } j \in C - B \\ \gamma_{g(j)}(x) \square \gamma_{h(j)}(x) & \text{if } j \in B \cap C \end{cases}$$

$$\delta_{u(j)}(x) = \begin{cases} \delta_{g(j)}(x) & \text{if } j \in B - C \\ \delta_{h(j)}(x) & \text{if } j \in C - B \\ \delta_{g(j)}(x) \wedge \delta_{h(j)}(x) & \text{if } j \in B \cap C \end{cases}$$

Let $(f, A) \cup ((g, B) \cup (h, C)) = (f, A) \cup (u, J) = (v, K)$, where $K = A \cup B \cup C$. Then for $x \in U$ and $k \in K$, we have

$$\mu_{v(k)}(x) = \begin{cases} \mu_{f(k)}(x) & \text{if } k \in A - B - C \\ \mu_{g(k)}(x) & \text{if } k \in B - C - A \\ \mu_{h(k)}(x) & \text{if } k \in C - A - B \\ \mu_{f(k)}(x) \vee \mu_{g(k)}(x) & \text{if } k \in (A \cap B) - C \\ \mu_{g(k)}(x) \vee \mu_{h(k)}(x) & \text{if } k \in (B \cap C) - A \\ \mu_{h(k)}(x) \vee \mu_{f(k)}(x) & \text{if } k \in (C \cap A) - B \\ \mu_{f(k)}(x) \vee \mu_{g(k)}(x) \vee \mu_{h(k)}(x) & \text{if } k \in A \cap B \cap C \end{cases}$$

$$\gamma_{v(k)}(x) = \begin{cases} \gamma_{f(k)}(x) & \text{if } k \in A - B - C \\ \gamma_{g(k)}(x) & \text{if } k \in B - C - A \\ \gamma_{h(k)}(x) & \text{if } k \in C - A - B \\ \gamma_{f(k)}(x) \square \gamma_{g(k)}(x) & \text{if } k \in (A \cap B) - C \\ \gamma_{g(k)}(x) \square \gamma_{h(k)}(x) & \text{if } k \in (B \cap C) - A \\ \gamma_{h(k)}(x) \square \gamma_{f(k)}(x) & \text{if } k \in (C \cap A) - B \\ \gamma_{f(k)}(x) \square \gamma_{g(k)}(x) \square \gamma_{h(k)}(x) & \text{if } k \in A \cap B \cap C \end{cases}$$

$$\delta_{v(k)}(x) = \begin{cases} \delta_{f(k)}(x) & \text{if } k \in A - B - C \\ \delta_{g(k)}(x) & \text{if } k \in B - C - A \\ \delta_{h(k)}(x) & \text{if } k \in C - A - B \\ \delta_{f(k)}(x) \wedge \delta_{g(k)}(x) & \text{if } k \in (A \cap B) - C \\ \delta_{g(k)}(x) \wedge \delta_{h(k)}(x) & \text{if } k \in (B \cap C) - A \\ \delta_{h(k)}(x) \wedge \delta_{f(k)}(x) & \text{if } k \in (C \cap A) - B \\ \delta_{f(k)}(x) \wedge \delta_{g(k)}(x) \wedge \delta_{h(k)}(x) & \text{if } k \in A \cap B \cap C \end{cases}$$

Consequently $((f, A) \cup (g, B)) \cup (h, C) = (f, A) \cup ((g, B) \cup (h, C))$. (6)–(8) can be proved similarly. $\square$

7.3 Interval-Valued Neutrosophic Soft Set Relations

In this section, we introduce the concept of IVNSS relations and study their basic properties. We also study various types of interval-valued intuitionistic fuzzy soft set relations.

Definition 7.14 Let U be an initial universe and (F, A) and (G, B) be two IVNSS. Then, a relation ρ between them is defined as a pair $(H, A \times B)$, where H is the mapping given by $H : A \times B \to \text{IVNS}^U$. This is called IVNSS-*relation*. The collection of relations on IVNSS on $A \times B$ over U is denoted by $\rho_U^N(A \times B)$.

Remark 7.15 Let U be an initial universe and $(F_1, A_1), (F_2, A_2), \ldots, (F_n, A_n)$ be n numbers of IVNSS over U. Then a relation ρ between them is defined as a pair $(H, A_1 \times A_2 \times \cdots \times A_n)$, where H is the mapping given by $H: A_1 \times A_2 \times \cdots \times A_n \to \text{IVNS}^U$.

Example 7.16 Let us consider an IVNSS (F, A) which describes the 'attractiveness of the houses' under consideration. Let, the universe set be $U = \{h_1, h_2, h_3, h_4, h_5\}$ and the set of parameter be $A = \{\text{beautiful}(e_1), \text{in green surrounding}(e_3)\}$.

Then, the tabular representation of the IVNSS (F, A) is given below:

U	e_1	e_3
h_1	([0.2, 0.6], [0.3, 0.4], [0.4, 0.8])	([0.5, 0.7], [0.2, 0.4], [0.3, 0.6])
h_2	([0.4, 0.7], [0.2, 0.3], [0.1, 0.2])	([0.2, 0.4], [0.5, 0.6], [0.7, 0.9])
h_3	([0.6, 0.8], [0.4, 0.7], [0.3, 0.4])	([0.1, 0.4], [0.2, 0.5], [0.5, 0.7])
h_4	([0.1, 0.3], [0.5, 0.7], [0.2, 0.5])	([0.4, 0.6], [0.7, 0.8], [0.1, 0.2])
h_5	([0.3, 0.5], [0.6, 0.8], [0.5, 0.6])	([0.5, 0.7], [0.3, 0.6], [0.3, 0.6])

Now let us consider the IVNSS (G, B) which describes the 'cost of the houses' under consideration. Let the universe set be $U = \{h_1, h_2, h_3, h_4, h_5\}$ and the set of parameter be $B = \{\text{costly}(e_2), \text{moderate}(e_4)\}$.

Then, the tabular representation of the IVNSS (G, B) is given below:

U	e_2	e_4
h_1	([0.5, 0.6], [0.2, 0.4], [0.3, 0.7])	([0.1, 0.2], [0.4, 0.5], [0.6, 0.8])
h_2	([0.6, 0.7], [0.1, 0.2], [0.4, 0.6])	([0.5, 0.7], [0.2, 0.3], [0.3, 0.4])
h_3	([0.3, 0.4], [0.4, 0.6], [0.5, 0.7])	([0.3, 0.4], [0.1, 0.4], [0.4, 0.6])
h_4	([0.6, 0.9], [0.5, 0.7], [0.2, 0.4])	([0.2, 0.5], [0.3, 0.5], [0.5, 0.7])
h_5	([0.2, 0.4], [0.3, 0.5], [0.1, 0.2])	([0.4, 0.5], [0.6, 0.9], [0.3, 0.7])

Then, a relation on the two given IVNSS are given below: $P = (H, A \times B)$:

U	(e_1, e_2)	(e_2, e_4)	(e_3, e_2)	(e_3, e_4)
h_1	([0.2, 0.4], [0.3, 0.5], [0.6, 0.8])	([0.2, 0.3], [0.5, 0.6], [0.7, 0.8])	([0.5, 0.8], [0.1, 0.2], [0.3, 0.4])	([0.1, 0.3], [0.4, 0.6], [0.1, 0.2])
h_2	([0.5, 0.7], [0.2, 0.4], [0.3, 0.6])	([0.7, 0.9], [0.2, 0.3], [0.1, 0.2])	([0.2, 0.3], [0.6, 0.8], [0.5, 0.7])	([0.3, 0.6], [0.2, 0.4], [0.5, 0.9])
h_3	([0.4, 0.6], [0.1, 0.3], [0.4, 0.5])	([0.5, 0.7], [0.4, 0.6], [0.2, 0.3])	([0.4, 0.6], [0.3, 0.5], [0.7, 0.8])	([0.4, 0.7], [0.1, 0.3], [0.3, 0.6])
h_4	([0.6, 0.8], [0.4, 0.6], [0.1, 0.2])	([0.4, 0.5], [0.1, 0.3], [0.4, 0.7])	([0.4, 0.7], [0.2, 0.3], [0.4, 0.5])	([0.2, 0.3], [0.5, 0.6], [0.7, 0.8])
h_5	([0.5, 0.9], [0.5, 0.6], [0.2, 0.4])	([0.2, 0.3], [0.4, 0.6], [0.2, 0.5])	([0.1, 0.4], [0.5, 0.6], [0.1, 0.4])	([0.5, 0.8], [0.1, 0.2], [0.3, 0.5])

The tabular representations of P is called *relational matrices* for P. From above, we have $\mu_{H(e_1,e_2)}(h_1) = [0.2, 0.4]$, $\gamma_{H(e_3,e_2)}(h_2) = [0.6, 0.8]$, $\delta_{H(e_3,e_4)}(h_4) = [0.7, 0.8]$, etc. But this intervals lies on the 1st row-1st column, 2nd row-3rd column and 4th row-4th column respectively. So, we denote $\mu_{H(e_1,e_2)}(h_1)|_{(1,1)} = [0.2, 0.4]$, $\gamma_{H(e_3,e_2)}(h_2)|_{(2,3)} = [0.6, 0.8]$, $\delta_{H(e_3,e_4)}(h_4)|_{(4,4)} = [0.7, 0.8]$, etc. to make the clear concept about what are the positions of the intervals in the relational matrices.

Definition 7.17 The *order* of the relational matrix is (α, β), where α = number of the universal points and β = number of pairs of parameters considered in the relational matrix.

In Example 7.16, the relational matrix for P is of order $(5, 4)$.

If $\alpha = \beta$, then the relation matrix is called a square matrix.

Definition 7.18 Let $P, Q \in \rho_U^N(A \times B)$ and the order of their relational matrices are same. Then we define

(i) $P \cup Q = (H \oplus J, A \times B)$ where $H \oplus J : A \times B \to \text{IVNS}^U$ is defined as $(H \oplus J)(e_i, e_j) = H(e_i) \cup J(e_j)$ for $(e_i, e_j) \in A \times B$.

(ii) $P \cap Q = (H \otimes J, A \times B)$ where $H \otimes J : A \times B \to \text{IVNS}^U$ is defined as $(H \otimes J)(e_i, e_j) = H(e_i) \cap J(e_j)$ for $(e_i, e_j) \in A \times B$.

(iii) $P^c = (\sim H, A \times B)$ where $\sim H : A \times B \to \text{IVNS}^U$ is defined as $\sim H(e_i, e_j) = [H(e_i, e_j)]^c$ for $(e_i, e_j) \in A \times B$.

Result 7.19 Let $P, Q, R \in \rho_U^N(A \times B)$ and the order of their relational matrices are same. Then, the following properties hold:

1. $(P \cup Q)^c = P^c \cap Q^c$
2. $(P \cap Q)^c = P^c \cup Q^c$
3. $P \cup (Q \cup R) = (P \cup Q) \cup R$
4. $P \cap (Q \cap R) = (P \cap Q) \cap R$
5. $P \cup (Q \cap R) = (P \cup Q) \cap (P \cup R)$
6. $P \cap (Q \cup R) = (P \cap Q) \cup (P \cap R)$

Definition 7.20 Let $P, Q \in \rho_U^N(A \times B)$ and the order of their relational matrices are same. Then, $P \subseteq Q$ iff $H(e_i, e_j) \subseteq J(e_i, e_j)$ for $(e_i, e_j) \in A \times B$, where $P = (H, A \times B)$ and $Q = (J, A \times B)$.

Definition 7.21 Let $P \in \rho_U^N(A \times B)$ and $P = (H, A \times B)$ whose relational matrix is a square matrix. Then, P is called a *reflexive* IVNSS-relation if for $(e_i, e_j) \in A \times B$ and $h_k \in U$ we have, $\mu_{H(e_i,e_j)}(h_k)|_{(m,n)} = [1, 1]$, $\gamma_{H(e_i,e_j)}(h_k)|_{(m,n)} = [0, 0]$, $\delta_{H(e_i,e_j)}(h_k)|_{(m,n)} = [0, 0]$ for $m = n = k$.

Example 7.22 Let $U = \{h_1, h_2, h_3, h_4\}$. Let us consider the IVNSSs (F, A) and (G, B) where $A = \{e_1, e_3\}$ and $B = \{e_2, e_4\}$. Then a reflexive IVNSS-relation between them is given as follows.

U	(e_1, e_2)	(e_1, e_4)	(e_3, e_2)	(e_3, e_4)
h_1	([1, 1], [0, 0], [0, 0])	([0.2, 0.3], [0.5, 0.6], [0.7, 0.8])	([0.5, 0.8], [0.1, 0.2], [0.3, 0.4])	([0.1, 0.3], [0.4, 0.6], [0.1, 0.2])
h_2	([0.5, 0.7], [0.2, 0.4], [0.3, 0.6])	([1, 1], [0, 0], [0, 0])	([0.2, 0.3], [0.6, 0.8], [0.5, 0.7])	([0.3, 0.6], [0.2, 0.4], [0.5, 0.9])
h_3	([0.4, 0.6], [0.1, 0.3], [0.4, 0.5])	([0.5, 0.7], [0.4, 0.6], [0.2, 0.3])	([1, 1], [0, 0], [0, 0])	([0.4, 0.7], [0.1, 0.3], [0.3, 0.6])
h_4	([0.6, 0.8], [0.4, 0.6], [0.1, 0.2])	([0.4, 0.5], [0.1, 0.3], [0.4, 0.7])	([0.4, 0.7], [0.2, 0.3], [0.4, 0.5])	([1, 1], [0, 0], [0, 0])

Definition 7.23 Let $P \in \rho_U^N(A \times B)$ and $P = (H, A \times B)$ whose relational matrix is a square matrix. Then P is called an *antireflexive* IVNSS-relation if for $(e_i, e_j) \in A \times B$ and $h_k \in U$ we have $\mu_{H(e_i,e_j)}(h_k)|_{(m,n)} = [0,0], \gamma_{H(e_i,e_j)}(h_k)|_{(m,n)} = [1,1], \delta_{H(e_i,e_j)}(h_k)|_{(m,n)} = [1,1]$ for $m = n = k$.

Example 7.24 Let $U = \{h_1, h_2, h_3, h_4\}$. Let us consider the IVNSS (F, A) and (G, B), where $A = \{e_1, e_3\}$ and $B = \{e_2, e_4\}$. Then, an antireflexive IVNSS-relation between them is given as follows:

U	(e_1, e_2)	(e_1, e_4)	(e_3, e_2)	(e_3, e_4)
h_1	([0, 0], [1,1] [1,1])	([0.2, 0.3], [0.5, 0.6], [0.7, 0.8])	([0.5, 0.8], [0.1, 0.2], [0.3, 0.4])	([0.1, 0.3], [0.4, 0.6], [0.1, 0.2])
h_2	([0.5, 0.7], [0.2, 0.4], [0.3, 0.6])	([0, 0], [1,1] [1,1])	([0.2, 0.3], [0.6, 0.8], [0.5, 0.7])	([0.3, 0.6], [0.2, 0.4], [0.5, 0.9])
h_3	([0.4, 0.6], [0.1, 0.3], [0.4, 0.5])	([0.5, 0.7], [0.4, 0.6], [0.2, 0.3])	([0, 0], [1,1] [1,1])	([0.4, 0.7], [0.1, 0.3], [0.3, 0.6])
h_4	([0.6, 0.8], [0.4, 0.6], [0.1, 0.2])	([0.4, 0.5], [0.1, 0.3], [0.4, 0.7])	([0.4, 0.7], [0.2, 0.3], [0.4, 0.5])	[0, 0], [1,1] [1,1])

Definition 7.25 Let $P \in \rho_U^N(A \times B)$ and $P = (H, A \times B)$ whose relational matrix is a square matrix. Then P is called a *symmetric* IVNSS-relation if for each $(e_i, e_j) \in A \times B$ and $h_k \in U$, $\exists (e_p, e_q) \in A \times B$ and $h_l \in U$ such that

$$\mu_{H(e_i,e_j)}(h_k)|_{(m,n)} = \mu_{H(e_p,e_q)}(h_l)|_{(n,m)},$$

$$\gamma_{H(e_i,e_j)}(h_k)|_{(m,n)} = \gamma_{H(e_p,e_q)}(h_l)|_{(n,m)},$$

$$\delta_{H(e_i,e_j)}(h_k)|_{(m,n)} = \delta_{H(e_p,e_q)}(h_l)|_{(n,m)}.$$

Example 7.26 Let $U = \{h_1, h_2, h_3, h_4\}$. Let us consider the IVNSS be (F, A) and (G, B), where $A = \{e_1, e_3\}$ and $B = \{e_2, e_4\}$. Then, a symmetric IVNSS-relation between them is given as follows:

U	(e_1, e_2)	(e_1, e_4)	(e_3, e_2)	(e_3, e_4)
h_1	([0.2, 0.4], [0.3, 0.5], [0.6, 0.8])	([0.2, 0.3], [0.5, 0.6], [0.7, 0.8])	([0.5, 0.8], [0.1, 0.2], [0.3, 0.4])	([0.6, 0.8], [0.4, 0.6], [0.1, 0.2])
h_2	([0.2, 0.3], [0.5, 0.6], [0.7, 0.8])	([0.7, 0.9], [0.2, 0.3], [0.1, 0.2])	([0.2, 0.3], [0.6, 0.8], [0.5, 0.7])	([0.4, 0.5], [0.1, 0.3], [0.4, 0.7])
h_3	([0.5, 0.8], [0.1, 0.2], [0.3, 0.4])	([0.2, 0.3], [0.6, 0.8], [0.5, 0.7])	([0.4, 0.6], [0.3, 0.5], [0.7, 0.8])	([0.4, 0.7], [0.2, 0.3], [0.4, 0.5])
h_4	([0.6, 0.8], [0.4, 0.6], [0.1, 0.2])	([0.4, 0.5], [0.1, 0.3], [0.4, 0.7])	([0.4, 0.7], [0.2, 0.3], [0.4, 0.5])	([0.2, 0.3], [0.5, 0.6], [0.7, 0.8])

Definition 7.27 Let $P \in \rho_U^N (A \times B)$ and $P = (H, A \times B)$ whose relational matrix is a square matrix. Then P is called an *antisymmetric* IVNSS-relation if for $(e_i, e_j) \in A \times B$ and $h_k \in U$, $\exists (e_p, e_q) \in A \times B$ and $h_l \in U$ such that either

$$\mu_{H(e_i,e_j)}(h_k)|_{(m,n)} \neq \mu_{H(e_p,e_q)}(h_l)|_{(n,m)},$$

$$\gamma_{H(e_i,e_j)}(h_k)|_{(m,n)} \neq \gamma_{H(e_p,e_q)}(h_l)|_{(n,m)},$$

$$\delta_{H(e_i,e_j)}(h_k)|_{(m,n)} \neq \delta_{H(e_p,e_q)}(h_l)|_{(n,m)}$$

or

$$\mu_{H(e_i,e_j)}(h_k)|_{(m,n)} = \mu_{H(e_p,e_q)}(h_l)|_{(n,m)} = [0,0],$$

$$\gamma_{H(e_i,e_j)}(h_k)|_{(m,n)} = \gamma_{H(e_p,e_q)}(h_l)|_{(n,m)} = [1,1],$$

$$\delta_{H(e_i,e_j)}(h_k)|_{(m,n)} = \delta_{H(e_p,e_q)}(h_l)|_{(n,m)} = [1,1]$$

Example 7.28 Let $U = \{h_1, h_2, h_3, h_4\}$. Let us consider the IVNSS (F, A) and (G, B), where $A = \{e_1, e_3\}$ and $B = \{e_2, e_4\}$. Then, an antisymmetric IVNSS-relation between them is given as follows:

U	(e_1, e_2)	(e_1, e_4)	(e_3, e_2)	(e_3, e_4)
h_1	([0.2, 0.4], [0.3, 0.5], [0.6, 0.8])	([0.2, 0.3], [0.5, 0.6], [0.7, 0.8])	([0, 0], [1, 1], [1, 1])	([0.6, 0.8], [0.4, 0.6], [0.1, 0.2])
h_2	([0.1, 0.4], [0.3, 0.5], [0.6, 0.7])	([0.7, 0.9], [0.2, 0.3], [0.1, 0.2)	([0.2, 0.3], [0.6, 0.8], [0.5, 0.7])	([0, 0], [1, 1], [1, 1])
h_3	([0, 0], [1, 1], [1, 1])	([0.1, 0.4], [0.5, 0.6], [0.6, 0.8])	([0.4, 0.6], [0.3, 0.5], [0.7, 0.8])	([0.4, 0.7], [0.2, 0.3], [0.4, 0.5])
h_4	([0.5, 0.7], [0.3, 0.7], [0.2, 0.4])	([0, 0], [1, 1], [1, 1])	([0.6, 0.8], [0.4, 0.5], [0.1, 0.2])	([0.2, 0.3], [0.5, 0.6], [0.7, 0.8])

Definition 7.29 Let $P \in \rho_U^N (A \times B)$ and $P = (H, A \times B)$ whose relational matrix is a square matrix. Then, P is called a *perfectly antisymmetric* IVNSS-relation if for $(e_i, e_j) \in A \times B$ and $h_k \in U$, $\exists (e_p, e_q) \in A \times B$ and $h_l \in U$ such that whenever

$$\inf \mu_{H(e_i,e_j)}(h_k)|_{(m,n)} > 0,\ \inf \gamma_{H(e_i,e_j)}(h_k)|_{(m,n)} > 0,\ \inf \delta_{H(e_i,e_j)}(h_k)|_{(m,n)} > 0,\ \text{then}$$

$$\inf \mu_{H(e_p,e_q)}(h_l)|_{(n,m)} = [0,0],\ \inf \gamma_{H(e_p,e_q)}(h_l)|_{(n,m)} = [1,1],\ \inf \delta_{H(e_p,e_q)}(h_l)|_{(n,m)}$$
$$= [1,1].$$

Example 7.30 Let $U = \{h_1, h_2, h_3, h_4\}$. Let us consider the IVNSS (F,A) and (G,B), where $A = \{e_1, e_3\}$ and $B = \{e_2, e_4\}$. Then a perfectly antisymmetric IVNSS-relation between them is given as follows:

U	(e_1, e_2)	(e_1, e_4)	(e_3, e_2)	(e_3, e_4)
h_1	([0.2, 0.4], [0.3, 0.5], [0.6, 0.8])	([0, 0], [1, 1], [1, 1])	([0.5, 0.8], [0.1, 0.2], [0.3, 0.4])	([0, 0], [1, 1], [1, 1])
h_2	([0.5, 0.7], [0.2, 0.4], [0.3, 0.6])	([0.7, 0.9], [0.2, 0.3], [0.1, 0.2)	([0, 0], [1, 1], [1, 1])	([0.3, 0.6], [0.2, 0.4], [0.5, 0.9])
h_3	([0, 0], [1, 1], [1, 1])	([0.1, 0.4], [0.5, 0.6], [0.6, 0.8])	([0.4, 0.6], [0.3, 0.5], [0.7, 0.8])	([0, 0], [1, 1], [1, 1])
h_4	([0.6, 0.8], [0.4, 0.6], [0.1, 0.2])	([0, 0], [1, 1], [1, 1])	([0.6, 0.8], [0.4, 0.5], [0.1, 0.2])	([0.2, 0.3], [0.5, 0.6], [0.7, 0.8])

Definition 7.31 Let $P, Q \in \rho_U^N(A \times A)$ and $P = (H, A \times A)$, $Q = (J, A \times A)$ and the order of their relational matrices are same. Then the composition of P and Q, denoted by $P * Q$, is defined by $P * Q = (H \circ J, A \times A)$, where $H \circ J : A \times A \to \text{IVNS}^U$ is defined as

$$(H \circ J)(e_i, e_j) = \left\{ \left\langle h_k, \mu_{(H \circ J)(e_i, e_j)}(h_k), \gamma_{(H \circ J)(e_i, e_j)}(h_k) \right\rangle : h_k \in U \right\},$$

where

$$\mu_{(H \circ J)(e_i, e_j)}(h_k) = \left[\max_l \left(\min \left(\inf \mu_{H(e_i, e_l)}(h_k), \inf \mu_{J(e_l, e_j)}(h_k) \right) \right), \right.$$
$$\left. \max_l \left(\min \left(\sup \mu_{H(e_i, e_l)}(h_k), \sup \mu_{J(e_l, e_j)}(h_k) \right) \right) \right]$$

$$\gamma_{(H \circ J)(e_i, e_j)}(h_k) = \left[\min_l \left(\max \left(\inf \gamma_{H(e_i, e_l)}(h_k), \inf \gamma_{J(e_l, e_j)}(h_k) \right) \right), \right.$$
$$\left. \min_l \left(\max \left(\sup \gamma_{H(e_i, e_l)}(h_k), \sup \gamma_{J(e_l, e_j)}(h_k) \right) \right) \right]$$

$$\delta_{(H \circ J)(e_i, e_j)}(h_k) = \left[\min_l \left(\max \left(\inf \delta_{H(e_i, e_l)}(h_k), \inf \delta_{J(e_l, e_j)}(h_k) \right) \right), \right.$$
$$\left. \min_l \left(\max \left(\sup \delta_{H(e_i, e_l)}(h_k), \sup \delta_{J(e_l, e_j)}(h_k) \right) \right) \right]$$

for $(e_i, e_j) \in A \times A$.

Example 7.32 Let $U = \{h_1, h_2, h_3, h_4\}$. Let us consider the IVNSS (F, A) and (G, B), where $A = \{e_1, e_2\}$. Let $P, Q \in \rho_U^N(A \times A)$ and $P = (H, A \times A)$, $Q = (J, A \times A)$, where

P:

U	(e_1, e_1)	(e_1, e_2)	(e_2, e_1)	(e_2, e_2)
h_1	([0.3, 0.4], [0.3, 0.4], [0.3, 0.4])	([0.2, 0.4], [0.3, 0.5], [0.2, 0.4])	([0.2, 0.5], [0.3, 0.4], [0.2, 0.5])	([0.2, 0.3], [0.3, 0.6], [0.3, 0.6])
h_2	([1, 1], [0, 0], [0, 0])	([0.1, 0.2], [0, 0], [0.1, 0.2])	([0.4, 0.5], [0.1, 0.3], [0.4, 0.5])	([0.4, 0.7], [0.1, 0.3], [0.1, 0.3])
h_3	([0.2, 0.6], [0.1, 0.4], [0.1, 0.4])	([0.2, 0.6], [0.1, 0.3], [0.2, 0.6])	([0.2, 0.3], [0.4, 0.6], [0.2, 0.3])	([0.2, 0.5], [0.2, 0.3], [0.2, 0.3])
h_4	([0.2, 0.4], [0.3, 0.5], [0.3, 0.5])	([0.3, 0.4], [0.4, 0.5], [0.3, 0.4])	([0.3, 0.4], [0.2, 0.3], [0.3, 0.4])	([0, 0.2], [0.4, 0.5], [0.4, 0.5])

Q:

U	(e_1, e_1)	(e_1, e_2)	(e_2, e_1)	(e_2, e_2)
h_1	([0.5, 0.8], [0.1, 0.2], [0.1, 0.2])	([0.2, 0.3], [0.3, 0.6], [0.2, 0.3])	([0.1, 0.4], [0.3, 0.5], [0.1, 0.4])	([0.2, 0.4], [0.2, 0.3], [0.2, 0.3])
h_2	([0.4, 0.5], [0.2, 0.4], [0.2, 0.4])	([0.4, 0.6], [0.2, 0.3], [0.4, 0.6])	([0.1, 0.5], [0.4, 0.5], [0.1, 0.5])	([0.4, 0.5], [0.1, 0.2], [0.1, 0.2])
h_3	([0.2, 0.3], [0.5, 0.6], [0.5, 0.6])	([0.3, 0.4], [0.4, 0.5], [0.3, 0.4])	([0.7, 0.8], [0.1, 0.2], [0.7, 0.8])	([0.3, 0.5], [0.3, 0.4], [0.3, 0.4])
h_4	([0.3, 0.5], [0.3, 0.4], [0.3, 0.4])	([0.3, 0.5], [0.2, 0.4], [0.3, 0.5])	([0.2, 0.4], [0.2, 0.3], [0.2, 0.4])	([0.3, 0.7], [0.1, 0.3], [0.1, 0.3])

Then, $P * Q$:

U	(e_1, e_1)	(e_1, e_2)	(e_2, e_1)	(e_2, e_2)
h_1	([0.3, 0.4], [0.3, 0.4], [0.3, 0.4])	([0.2, 0.4], [0.3, 0.5], [0.2, 0.4])	([0.1, 0.5], [0.3, 0.4], [0.2, 0.5])	([0.2, 0.3], [0.3, 0.6], [0.3, 0.6])
h_2	([0.4, 0.5], [0.2, 0.4], [0.2, 0.4])	([0.1, 0.6], [0.1, 0.2], [0.1, 0.6])	([0.4, 0.5], [0.2, 0.4], [0.4, 0.5])	([0.4, 0.5], [0.1, 0.3], [0.1, 0.3])
h_3	([0.2, 0.6], [0.1, 0.3], [0.1, 0.3])	([0.2, 0.5], [0.3, 0.4], [0.2, 0.5])	([0.2, 0.5], [0.2, 0.3], [0.2, 0.5])	([0.2, 0.5], [0.3, 0.4], [0.3, 0.4])
h_4	([0.2, 0.4], [0.3, 0.5], [0.3, 0.5])	([0.3, 0.4], [0.3, 0.5], [0.3, 0.4])	([0.3, 0.4], [0.3, 0.4], [0.3, 0.4])	([0.3, 0.4], [0.2, 0.3], [0.2, 0.4])

Definition 7.33 Let $P \in \rho_U^N(A \times A)$ and $P = (H, A \times A)$. Then P is called a *transitive* IVNSS-relation if $P * P \subseteq P$.

Example 7.34 Let $U = \{h_1, h_2, h_3, h_4\}$. Let us consider the IVNSS (F, A), where $A = \{e_1, e_2\}$. Let $P \in \rho_U^N (A \times A)$ and $P = (H, A \times A)$, where

P:

U	(e_1, e_1)	(e_1, e_2)	(e_2, e_1)	(e_2, e_2)
h_1	([0.3, 0.4], [0.3, 0.4], [0.3, 0.4])	([0.2, 0.4], [0.3, 0.6], [0.2, 0.4])	([0.2, 0.5], [0.3, 0.4], [0.2, 0.5])	([0.2, 0.4], [0.3, 0.6], [0.3, 0.6])
h_2	([1,1], [0, 0], [0, 0])	([0.1, 0.2], [0, 0], [0.1, 0.2])	([0.4, 0.5], [0.1, 0.3], [0.4, 0.5])	([0.4, 0.7], [0.1, 0.3], [0.1, 0.3])
h_3	([0.2, 0.6], [0.1, 0.4], [0.1, 0.4])	([0.2, 0.6], [0.1, 0.3], [0.2, 0.6])	([0.2, 0.3], [0.4, 0.6], [0.2, 0.3])	([0.2, 0.5], [0.2, 0.3], [0.2, 0.3])
h_4	([0.3, 0.4], [0.3, 0.4], [0.3, 0.4])	([0.2, 0.4], [0.3, 0.5], [0.2, 0.4])	([0.2, 0.5], [0.3, 0.4], [0.2, 0.5])	([0.2, 0.4], [0.3, 0.5], [0.3, 0.5])

Then $P * P$:

U	(e_1, e_1)	(e_1, e_2)	(e_2, e_1)	(e_2, e_2)
h_1	([0.3, 0.4], [0.3, 0.4], [0.3, 0.4])	([0.2, 0.4], [0.3, 0.6], [0.2, 0.4])	([0.2, 0.4], [0.3, 0.4], [0.2, 0.4)	([0.2, 0.4], [0.3, 0.6], [0.3, 0.6])
h_2	([1,1], [0, 0], [0, 0])	([0.1, 0.2], [0, 0], [0.1, 0.2])	([0.4, 0.5], [0.1, 0.3], [0.4, 0.5])	([0.4, 0.7], [0.1, 0.3], [0.1, 0.3])
h_3	([0.2, 0.6], [0.1, 0.4], [0.1, 0.4])	([0.2, 0.6], [0.1, 0.3], [0.2, 0.6])	([0.2, 0.3], [0.4, 0.6], [0.2, 0.3])	([0.2, 0.5], [0.2, 0.3], [0.2, 0.3])
h_4	([0.3, 0.4], [0.3, 0.4], [0.3, 0.4])	([0.2, 0.4], [0.3, 0.5], [0.2, 0.4])	([0.2, 0.4], [0.3, 0.4], [0.2, 0.4])	([0.2, 0.4], [0.3, 0.5], [0.3, 0.5])

Thus, $P * P \subset P$ and so P ia a transitive IVIFSS- relation.

7.4 Solution of a Decision-Making Problem

The concept of interval-valued neutrosophic soft relations can be used effectively for solving a wide range of decision-making problems.

Example 7.35 Let $U = \{h_1, h_2, h_3, h_4, h_5\}$ be the set of five houses and $E = \{e_1(\text{expensive}), e_2(\text{wooden}), e_3(\text{beautiful}), e_4(\text{cheap}), e_5(\text{in the green surroundings}), e_6(\text{concrete}), e_7(\text{in the main town}), e_8(\text{moderate})\}$ be the set of parameters. Let us consider the four IVNSS $(F_1, A_1), (F_2, A_2), (F_3, A_3), (F_4, A_4)$ which describes the 'cost of the houses', 'attractiveness of the houses', 'physical trait of the houses', and 'characteristic of the place where the houses are located' respectively. Now suppose that Mr. X is interested in buying a house on the basis of his choice of parameters *'beautiful', 'wooden', 'cheap',* and *'in the green surroundings'*. This implies that from the available houses in U, he will select the house which satisfies with all the parameters of his choice.

Step-1: Let the tabular representations of the above soft sets are given by, respectively:

(F_1, A_1):

U	e_1	e_4
h_1	([0.2, 0.6], [0.3, 0.4], [0.4, 0.8])	([0.5, 0.7], [0.2, 0.4], [0.3, 0.6])
h_2	([0.4, 0.7], [0.2, 0.3], [0.1, 0.2])	([0.2, 0.4], [0.5, 0.6], [0.7, 0.9])
h_3	([0.6, 0.8], [0.4, 0.7], [0.3, 0.4])	([0.1, 0.4], [0.2, 0.5], [0.5, 0.7])
h_4	([0.1, 0.3], [0.5, 0.7], [0.2, 0.5])	([0.4, 0.6], [0.7, 0.8], [0.1, 0.2])
h_5	([0.3, 0.5], [0.6, 0.8], [0.5, 0.6])	([0.5, 0.7], [0.3, 0.6], [0.3, 0.6])

(F_2, A_2):

U	e_3	e_8
h_1	([0.2, 0.3], [0.1, 0.4], [0.6, 0.9])	([0.4, 0.5], [0.6, 0.7], [0.1, 0.4])
h_2	([0.5, 0.6], [0.2, 0.5], [0.3, 0.4])	([0.3, 0.4], [0.2, 0.4], [0.6, 0.7])
h_3	([0.1, 0.3], [0.5, 0.6], [0.2, 0.4])	([0.5, 0.7], [0.4, 0.6], [0.2, 0.5])
h_4	([0.4, 0.6], [0.2, 0.3], [0.4, 0.5])	([0.1, 0.2], [0.3, 0.5], [0.5, 0.8])
h_5	([0.7, 0.8], [0.4, 0.5], [0.1, 0.2])	([0.6, 0.7], [0.4, 0.8], [0.2, 0.4])

(F_3, A_3):

U	e_2	e_6
h_1	([0.7, 0.8], [0.3, 0.5], [0.2, 0.3])	([0.2, 0.3], [0.3, 0.4], [0.2, 0.5])
h_2	([0.1, 0.3], [0.4, 0.6], [0.3, 0.6])	([0.6, 0.8], [0.1, 0.4], [0.2, 0.4])
h_3	([0.5, 0.6], [0.2, 0.4], [0.6, 0.7])	([0.3, 0.6], [0.3, 0.6], [0.4, 0.6])
h_4	([0.3, 0.5], [0.4, 0.5], [0.4, 0.7])	([0.1, 0.3], [0.5, 0.6], [0.3, 0.5])
h_5	([0.4, 0.6], [0.5, 0.8], [0.2, 0.5])	([0.2, 0.6], [0.2, 0.4], [0.6, 0.9])

(F_4, A_4):

U	e_5	e_7
h_1	([0.5, 0.6], [0.2, 0.5], [0.5, 0.7])	([0.6, 0.7], [0.2, 0.4], [0.1, 0.6])
h_2	([0.2, 0.4], [0.5, 0.7], [0.2, 0.5])	([0.2, 0.4], [0.3, 0.4], [0.5, 0.7])
h_3	([0.4, 0.5], [0.3, 0.4], [0.1, 0.2])	([0.5, 0.8], [0.5, 0.7], [0.2, 0.4])
h_4	([0.1, 0.5], [0.6, 0.9], [0.3, 0.5])	([0.4, 0.6], [0.3, 0.7], [0.7, 0.8])
h_5	([0.2, 0.6], [0.1, 0.3], [0.2, 0.3])	([0.3, 0.5], [0.2, 0.6], [0.4, 0.7])

Step-2: To solve this problem, let us consider an interval-valued neutrosophic soft set relation $\rho = (F, A_1 \times A_2 \times A_3 \times A_4)$, where $F : A_1 \times A_2 \times A_3 \times A_4 \rightarrow$ IVNS^U is defined by

$$F((e_i, e_j), (e_k, e_l)) = \left\{ \left\langle h_\lambda, \left[\max\left(\inf \mu_{F_1(e_i)}(h_\lambda), \inf \mu_{F_2(e_j)}(h_\lambda), \inf \mu_{F_3(e_k)}(h_\lambda), \right. \right. \right. \right.$$
$$\left. \inf \mu_{F_4(e_l)}(h_\lambda) \right), \max\left(\sup \mu_{F_1(e_i)}(h_\lambda), \sup \mu_{F_2(e_j)}(h_\lambda), \right.$$
$$\left. \left. \sup \mu_{F_3(e_k)}(h_\lambda), \sup \mu_{F_4(e_l)}(h_\lambda) \right) \right], \left[\min\left(\inf \gamma_{F_1(e_i)}(h_\lambda), \right. \right.$$
$$\left. \inf \gamma_{F_2(e_j)}(h_\lambda), \inf \gamma_{F_3(e_k)}(h_\lambda), \inf \gamma_{F_4(e_l)}(h_\lambda) \right),$$
$$\min\left(\sup \gamma_{F_1(e_i)}(h_\lambda), \sup \gamma_{F_2(e_j)}(h_\lambda), \sup \gamma_{F_3(e_k)}(h_\lambda), \right.$$
$$\left. \left. \sup \gamma_{F_4(e_l)}(h_\lambda) \right) \right], \left[\min\left(\inf \delta_{F_1(e_i)}(h_\lambda), \inf \delta_{F_2(e_j)}(h_\lambda), \right. \right.$$
$$\left. \inf \delta_{F_3(e_k)}(h_\lambda), \inf \delta_{F_4(e_l)}(h_\lambda) \right), \min\left(\sup \delta_{F_1(e_i)}(h_\lambda), \right.$$
$$\left. \left. \left. \left. \sup \delta_{F_2(e_j)}(h_\lambda), \sup \delta_{F_3(e_k)}(h_\lambda), \sup \delta_{F_4(e_l)}(h_\lambda) \right) \right] \right\rangle : h_\lambda \in U \right\}$$

Then we have

$$F((e_4, e_3), (e_2, e_5)) = \{ \langle h_1, [\max(0.5, 0.2, 0.7, 0.5), \max(0.7, 0.3, 0.8, 0.6)],$$
$$[\min(0.2, 0.1, 0.3, 0.2), \min(0.4, 0.4, 0.5, 0.5)],$$
$$[\min(0.3, 0.6, 0.2, 0.5), \min(0.6, 0.9, 0.3, 0.7)] \rangle,$$
$$\langle h_2, [\max(0.2, 0.5, 0.1, 0.2), \max(0.4, 0.6, 0.3, 0.4)],$$
$$[\min(0.5, 0.2, 0.4, 0.5), \min(0.6, 0.5, 0.6, 0.7)],$$
$$[\min(0.7, 0.3, 0.3, 0.2), \min(0.9, 0.4, 0.6, 0.5)] \rangle,$$
$$\langle h_3, [\max(0.1, 0.1, 0.5, 0.4), \max(0.4, 0.3, 0.6, 0.5)],$$
$$[\min(0.2, 0.5, 0.2, 0.3), \min(0.5, 0.6, 0.4, 0.4)],$$
$$[\min(0.5, 0.2, 0.6, 0.1), \min(0.7, 0.4, 0.7, 0.2)] \rangle,$$
$$\langle h_4, [\max(0.4, 0.4, 0.3, 0.1), \max(0.6, 0.6, 0.5, 0.5)],$$
$$[\min(0.7, 0.2, 0.4, 0.6), \min(0.8, 0.3, 0.5, 0.9)],$$
$$[\min(0.1, 0.4, 0.4, 0.3), \min(0.2, 0.5, 0.7, 0.5)] \rangle,$$
$$\langle h_5, [\max(0.5, 0.7, 0.4, 0.2), \max(0.7, 0.8, 0.6, 0.6)],$$
$$[\min(0.3, 0.4, 0.5, 0.1), \min(0.6, 0.5, 0.8, 0.3)],$$
$$[\min(0.3, 0.1, 0.2, 0.2), \min(0.6, 0.2, 0.5, 0.3)] \rangle)$$

i.e.

$$F((e_4, e_3), (e_2, e_5)) = \{\langle h_1, [0.7, 0.8], [0.1, 0.4], [0.2, 0.3]\rangle,$$
$$\langle h_2, [0.5, 0.6], [0.2, 0.5], [0.2, 0.4]\rangle,$$
$$\langle h_3, [0.5, 0.6], [0.2, 0.4], [0.1, 0.2]\rangle,$$
$$\langle h_4, [0.4, 0.6], [0.2, 0.3], [0.1, 0.2]\rangle,$$
$$\langle h_5, [0.7, 0.8], [0.1, 0.3], [0.1, 0.2]\rangle\}$$

Step-3: We define the score of h_i (denoted by $S(h_i)$) by $\forall h_i \in U$,

$$S(h_i) = 2 * \sup \mu(h_i) - (\sup \gamma(h_i) + \sup \delta(h_i)), \quad h_i \in U.$$

Then, we have $S(h_1) = 0.9, S(h_2) = 0.3, S(h_3) = 0.6, S(h_4) = 0.7, S(h_5) = 1.1$.
Step-4: Now we define the decision set as $D = \{h_i \in U : S(h_i) \text{ is maximum}\}$.
Then, we have $D = \{h_5\}$. So, Mr. X will buy the house h_5.

References

1. Ali, M.I., Feng, F., Liu, X., Min, W.K., Shabir, M.: On some new operations in soft set theory. Comput. Math. Appl. **57**(9), 1547–1553 (2009)
2. Atanassov, K.: Intuitionistic fuzzy sets. Fuzzy Sets Syst. **20**, 87–96 (1986)
3. Maji, P.K., Biswas, R., Roy, A.R.: Soft set theory. Comput. Math. Appl. **45**(4–5), 555–562 (2003)
4. Maji, P.K.: Neutrosophic soft set. Ann. Fuzzy Math. Inf. **5**(1), 157–168 (2013)
5. Molodtsov, D.: Soft set theory-first results. Comput. Math. Appl. **37**(4–5), 19–31 (1999)
6. Pawlak, Z.: Rough sets. Int. J. Comput. Inf. Sci. **11**, 341–356 (1982)
7. Smarandache, F.: Neutrosophic set-a generalisation of the intuitionistic fuzzy sets. Int. J. Pure Appl. Math. **24**, 287–297 (2005)
8. Turksen, I.: Interval valued fuzzy sets based on normal forms. Fuzzy Sets Syst. **20**, 191–210 (1986)
9. Zadeh, L.A.: Fuzzy sets. Inf. Control **8**, 338–353 (1965)
10. Zadeh, L.A.: The concept of a linguistic variable and its application to approximate reasoning-I. Inform. Sci. **8**, 199–249 (1975)

Chapter 8
Topological Structure Formed by Soft Multi-Sets and Soft Multi-Compact Spaces

Abstract The purpose of this chapter was to study the concept of topological structure formed by soft multi-sets. The notion of relative complement of soft multi-set, soft multi-point, soft multi-open set, soft multi-closed set, soft multi-basis, soft multi-sub-basis, neighbourhoods and neighbourhood system, interior and closure of a soft multi-set, etc., is to be introduced, and their basic properties are also to be investigated. It is seen that a soft multi-topological space gives a parameterised family of topological spaces. Lastly, the concept of soft multi-compact space is also introduced.

Keywords Soft multi-set · Soft multi-topology · Soft multi-point · Soft multi-open set · Soft multi-closed set · Neighbourhood of a soft multi-set · Interior of a soft multi-set · Closure of a soft multi-set · Soft multi-basis · Soft multi-compact space

In 1999, Molodtsov initiated soft set theory as a completely generic mathematical tool for modelling vague concepts. In soft set theory, there is no limited condition to the description of objects; so researchers can choose the form of parameters as they need, which greatly simplifies the decision-making process and makes the process more efficient in the absence of partial information. Although many mathematical tools are available for modelling uncertainties such as probability theory, fuzzy set theory, rough set theory, and interval-valued mathematics, there are inherent difficulties associated with each of these techniques. Moreover, all these techniques lack in the parameterisation of the tools, and hence, they could not be applied successfully in tackling problems especially in areas such as economic, environmental, and social problems domains. Soft set theory is standing in a unique way in the sense that it is free from the above difficulties.

Soft set theory has a rich potential for application in many directions, some of which are reported by Molodtsov [7] in his work. Later on, Maji et al. [5, 6] presented some new definitions on soft sets such as subset, union, intersection, and complements of soft sets and discussed in detail the application of soft set in decision-making problem. Based on the analysis of several operations on soft sets introduced in [6], Ali et al. [4] presented some new algebraic operations for soft sets

© Springer India 2015
A. Mukherjee, *Generalized Rough Sets*, Studies in Fuzziness
and Soft Computing 324, DOI 10.1007/978-81-322-2458-7_8

and proved that certain De Morgan's law holds in soft set theory with respect to these new definitions. Alkhazaleh et al. [1, 2] as a generalisation of Molodtsov's soft set presented the definition of a soft multi-set and its basic operations such as complement, union, and intersection Recently, Mukherjee and Das [8, 9] introduced soft multi-topology. In this chapter, we study the concept of soft multi-topological spaces in details. Then, the notion of relative complement of soft multi-set, soft multi-point, soft multi-open set, soft multi-closed set, soft multi-basis, soft multi-sub-basis, neighbourhood and neighbourhood system, interior and closure of a soft multi-set, etc., is to be introduced, and their basic properties are investigated. It is seen that a soft multi-topological space gives a parameterised family of topological spaces. Lastly, we introduce the soft multi-compact space.

Definition 8.1 [1, 3] Let $\{U_i : i \in I\}$ be a collection of universes such that $\cap_{i \in I} U_i = \phi$ and let $\{E_{U_i} : i \in I\}$ be a collection of sets of parameters. Let $U = \prod_{i \in I} P(U_i)$, where $P(U_i)$ denotes the power set of U_i, $E = \prod_{i \in I} E_{U_i}$, and $A \subseteq E$. A pair (F, A) is called a soft multi-set over U, where F is a mapping given by $F : A \to U$.

Definition 8.2 [1] A soft multi-set (F, A) over U is called a null soft multi-set denoted by $\tilde{\phi}$, if for all $a \in A, F(a) = \phi$.

Definition 8.3 [1] A soft multi-set (F, A) over U is called an absolute soft multi-set denoted by $\tilde{U}$, if for all $a \in A, F(a) = U$.

Definition 8.4 [1] For any soft multi-set (F, A), a pair $\left(e_{U_{i,j}}, F_{e_{U_{i,j}}}\right)$ is called a U_i— soft multi-part $\forall e_{U_{i,j}} \in a_k$ and $F_{e_{U_{i,j}}} \subseteq F(A)$ are an approximate value set, where $a_k \in A, k = \{1, 2, 3, \ldots, n\}, i \in \{1, 2, 3, \ldots, m\}$, and $j \in \{1, 2, 3, \ldots, r\}$.

Definition 8.5 [1] For two soft multi-sets (F, A) and (G, B) over U, (F, A) is called a soft multi-subset of (G, B) if

1. $A \subseteq B$ and
2. $\forall e_{U_{i,j}} \in a_k, \left(e_{U_{i,j}}, F_{e_{U_{i,j}}}\right) \subseteq \left(e_{U_{i,j}}, G_{e_{U_{i,j}}}\right)$

where $a_k \in A, k = \{1, 2, 3, \ldots, n\}, i \in \{1, 2, 3, \ldots, m\}$, and $j \in \{1, 2, 3, \ldots, r\}$. This relation is denoted by $(F, A) \tilde{\subseteq} (G, B)$. In this case, (G, B) is called a soft multi-superset of (F, A).

Definition 8.6 [1] Two soft multi-sets (F, A) and (G, B) over U are said to be equal if (F, A) is a soft multi-subset of (G, B) and (G, B) is a soft multi-subset of (F, A).

Definition 8.7 [1] Union of two soft multi-sets (F, A) and (G, B) over U denoted by $(F, A) \tilde{\cup} (G, B)$ is the soft multi-set (H, C), where $C = A \cup B$ and $\forall e \in C$,

$$H(e) = \begin{cases} F(e), & \text{if } e \in A - B \\ G(e), & \text{if } e \in B - A. \\ F(e) \cup G(e), & \text{if } e \in A \cap B \end{cases}$$

Definition 8.8 [1] Intersection of two soft multi-sets (F,A) and (G,B) over U denoted by $(F,A) \, \tilde{\cap} \, (G,B)$ is the soft multi-set (H, C), where $C = A \cup B$ and $\forall e \in C$,

$$H(e) = \begin{cases} F(e), & \text{if } e \in A - B \\ G(e), & \text{if } e \in B - A. \\ F(e) \cap G(e), & \text{if } e \in A \cap B \end{cases}$$

Definition 8.9 [1] Let $E = \prod_{i=1}^{m} E_{U_i}$, where E_{U_i} is a set of parameter of U_i. The NOT set of E denoted by $\neg E$ is defined by

$$\neg E = \prod_{i=1}^{m} \neg E_{U_i} \text{ where } \neg E_{U_i} = \{\neg e_{U_i,j} = \text{not } e_{U_i,j}, \forall i,j\}.$$

Definition 8.10 [1] The complement of a soft multi-set (F,A) over U is denoted by $(F,A)^c$ and is defined by $(F,A)^c = (F^c, \neg A)$, where $F^c : \neg A \to U$ is a mapping given by $F^c(\alpha) = U - F(\neg \alpha), \forall \alpha \in \neg A$.

8.1 Soft Multi-Sets and Their Basic Properties

In this section, for the sake of simplicity, we restate a few basic definitions, e.g. Definitions 8.3 and 8.4 in the following form to study a few results of soft multi-topological spaces properly. Let $\{U_i : i \in I\}$ be a collection of universes such that $\cap_{i \in I} U_i = \phi$ and let $\{E_{U_i} : i \in I\}$ be a collection of sets of parameters. Let $U = \prod_{i \in I} P(U_i)$, where $P(U_i)$ denotes the power set of U_i, $E = \prod_{i \in I} E_{U_i}$. The set of all soft multi-set over (U,E) is denoted by SMS(U,E).

Definition 8.11 A soft multi-set $(F,E) \in$ SMS(U,E) is called a null soft multi-set denoted by $\tilde{\phi}$, if for all $e \in E, F(e) = \phi$.

Definition 8.12 A soft multi-set $(F,E) \in$ SMS(U,E) is called an absolute soft multi-set denoted by $\tilde{E}$, if for all $e \in E, F(e) = U$.

Definition 8.13 The relative complement of a soft multi-set (F,A) over (U,E) is denoted by $(F,A)'$ and is defined by $(F,A)' = (F', A)$, where $F' : A \to U$ is a mapping given by $F'(e) = U - F(e), \forall e \in E$.

Example 8.14 Let us consider there are three universes U_1, U_2, and U_3.
  Let $U_1 = \{h_1, h_2, h_3, h_4\}$, $U_2 = \{c_1, c_2, c_3\}$ and $U_3 = \{v_1, v_2\}$. Let $\{E_{U_1}, E_{U_2}, E_{U_3}\}$ be a collection of sets of decision parameters related to the above universes, where

$$E_{U_1} = \{e_{U_1,1}, e_{U_1,2}, e_{U_1,3}\}, \quad E_{U_2} = \{e_{U_2,1}, e_{U_2,2}, e_{U_2,3}\},$$
$$E_{U_3} = \{e_{U_3,1}, e_{U_3,2}, e_{U_3,3}\}.$$

Let $U = \prod_{i=1}^{3} P(U_i)$, $E = \prod_{i=1}^{3} E_{U_i}$, and

$$A = \{e_1 = (e_{U_1,1}, e_{U_2,1}, e_{U_3,1}), e_2 = (e_{U_1,1}, e_{U_2,2}, e_{U_3,1})\}$$

Then, the relative complement of the soft multi-set

$$(F, A) = \{(e_1, (\{h_1, h_2\}, \{c_1, c_2\}, \{v_1\})), (e_2, (\{h_3, h_4\}, \{c_1, c_3\}, \{v_2\}))\},$$
is $(F, A)' = \{(e_1, (\{h_3, h_4\}, \{c_3\}, \{v_2\})), (e_2, (\{h_1, h_2\}, \{c_2\}, \{v_1\}))\}.$

Clearly, $\tilde{\phi}' = \tilde{E}$ and $\tilde{E}' = \tilde{\phi}$.

Proposition 8.15 *If (F, A) and (G, B) are two soft multi-sets over (U, E), then we have the following*

    (i) $((F, A) \ \tilde{\cup} \ (G, B))' = (F, A)' \ \tilde{\cap} \ (G, B)'$
    (ii) $((F, A) \ \tilde{\cap} \ (G, B))' = (F, A)' \ \tilde{\cup} \ (G, B)'$

Proof Straightforward.   □

Definition 8.16 A soft multi-set $(F, A) \in \text{SMS}(U, E)$ is called a soft multi-point in (U, E), denoted by $e_{(F,A)}$, if for the element $e \in A, F(e) \neq \varphi$, and $\forall e' \in A - \{e\}$, $F(e') = \varphi$.

Example 8.17 Let us consider there are three universes U_1, U_2, and U_3.
    Let $U_1 = \{h_1, h_2, h_3, h_4\}$, $U_2 = \{c_1, c_2, c_3\}$ and $U_3 = \{v_1, v_2\}$.   Let $\{E_{U_1}, E_{U_2}, E_{U_3}\}$ be a collection of sets of decision parameters related to the above universes, where

$$E_{U_1} = \{e_{U_1,1} = \text{expensive}, e_{U_1,2} = \text{cheap}, e_{U_1,3} = \text{wooden}\}$$
$$E_{U_2} = \{e_{U_2,1} = \text{expensive}, e_{U_2,2} = \text{cheap}, e_{U_2,3} = \text{sporty}\},$$
$$E_{U_3} = \{e_{U_3,1} = \text{expensive}, e_{U_3,2} = \text{cheap}, e_{U_3,3} = \text{in Kuala Lumpur}\}.$$

Let $U = \prod_{i=1}^{3} P(U_i)$, $E = \prod_{i=1}^{3} E_{U_i}$ and

$$A = \{e_1 = (e_{U_1,1}, e_{U_2,1}, e_{U_3,1}), e_2 = (e_{U_1,1}, e_{U_2,2}, e_{U_3,1}), e_3 = (e_{U_1,2}, e_{U_2,3}, e_{U_3,1})\}$$

Then, the soft multi-set $(F, A) = \{(e_1, (\{h_1, h_2\}, \{c_1, c_2\}, \phi))\}$ is the soft multi-point, and it is denoted by $e_{1_{(F,A)}}$.

Definition 8.18 A soft multi-point $e_{(F,A)}$ is said to be in the soft multi-set (G,B), denoted by $e_{(F,A)} \, \tilde{\in} \, (G,B)$, if $(F,A) \, \tilde{\subseteq} \, (G,B)$.

Example 8.19 The soft multi-point $e_{1_{(F,A)}}$, as in the Example 8.17, in the soft multi-set $(G,B) = \{(e_1,(\{h_1,h_2\},\{c_1,c_2\},\{v_1\})), (e_2,(\{h_3,h_4\},\{c_1,c_3\},\{v_2\})), (e_3,(\{h_1,h_3,h_4\}, \{c_1,c_3\},\{v_1,v_2\}))\}$, i.e. $e_{(F,A)} \, \tilde{\in} \, (G,B)$,

Proposition 8.20 *Let $e_{(F,A)}$ be a soft multi-point and (G,B) be the soft multi-set in* $SMS(U,E)$. *If $e_{(F,A)} \, \tilde{\in} \, (G,B)$, then $e_{(F,A)} \, \tilde{\notin} \, (G,B)^c$.*

Proof If $e_{(F,A)} \, \tilde{\in} \, (G,B)$, then $(F,A) \, \tilde{\subseteq} \, (G,B)$, i.e. for the element $e \in A, F(e) \subseteq G(e)$.

This implies $F(e) \not\subseteq U - G(e) = G^c(e)$, i.e. $(F,A) \, \tilde{\not\subseteq} \, (G,B)^c$. Therefore, we have $e_{(F,A)} \, \tilde{\notin} \, (G,B)^c$. $\qquad\qquad\qquad\qquad\qquad\qquad\qquad\qquad\qquad\qquad\square$

Remark 8.21 The converse of the above proposition is not true in general.

Example 8.22 If we consider the soft multi-point

$$e_{1_{(F,A)}} = \{(e_1,(\{h_1,h_2\},\{c_1,c_2\},\phi))\}$$

as in the Example 8.17 and a soft multi-set

$$(G,B) = \{(e_1,(\{h_1,h_3\},\{c_2,c_3\},\{v_1\})), (e_2,(\{h_2,h_4\},\{c_1,c_3\},\{v_2\})),$$
$$(e_3,(\{h_4\},\{c_1\},\{v_2\}))\},$$

then $e_{1_{(F,A)}} \, \tilde{\notin} \, (G,B)$ and also $e_{1_{(F,A)}} \, \tilde{\notin} \, (G,B)^c$

$$= \{(e_1,(\{h_2,h_4\},\{c_1\},\{v_2\})), (e_2,(\{h_1,h_3\},\{c_2\},\{v_1\})),$$
$$(e_3,(\{h_1,h_2,h_3\},\{c_2,c_3\},\{v_1\}))\}.$$

Definition 8.23 Let $(F,A) \in SMS(U,E)$ and $x \in U_i$, for some i. Then, we say that $x \in (F,A)$ and read as x belongs to the soft multi-set (F,A) if $x \in F_{e_{U_i,j}}, \forall j$.

Example 8.24 Let us consider the soft multi-set $(F,A) = \{(e_1,(\{h_1,h_2\},\{c_1,c_2\}, \{v_1\})), (e_2,(\{h_3,h_4\},\{c_1,c_3\},\{v_2\}))\}$, as in the Example 3.4, and then, for the element $c_1 \in U_2$, we say that $c_1 \in (F,A)$, since $c_1 \in F_{e_{U_2,1}} = \{c_1,c_2\}$ and $c_1 \in F_{e_{U_2,2}} = \{c_1,c_3\}$ but $h_1,h_2 \notin (F,A)$ since $h_1,h_2 \in F_{e_{U_1,1}} = \{h_1,h_2\}$ but $h_1,h_2 \notin F_{e_{U1,2}} = \{h_3,h_4\}$.

Remark 8.25 For any $x \in U_i$, we say that $x \notin (F,A)$ if $x \notin F(e_{U_i,j})$ for some $e_{U_i,j} \in a_k, a_k \in A$.

8.2 Soft Multi-Topological Spaces

Recently, Mukherjee and Das [8, 9] introduced soft multi-topology. In this section, we study that the notion of relative complement of soft multi-set, soft multi-point, soft multi-set topology, soft multi-closed set, soft multi-basis, soft multi-sub-basis, neighbourhood and neighbourhood system, interior and closure of a soft multi-set, etc., is to be introduced, and their basic properties are investigated. It is seen that a soft multi-topological space gives a parameterised family of topological spaces.

Definition 8.26 A subfamily τ of $SMS(U, E)$ is called soft multi-set topology on (U, E), if the following axioms are satisfied:

$[O_1]$ $\tilde{\phi}, \tilde{E} \in \tau$,
$[O_2]$ $\{(F^k, A^k) | k \in K\} \subseteq \tau \Rightarrow \tilde{U}_{k \in K}(F^k, A^k) \in \tau$,
$[O_3]$ If $(F, A), (G, B) \in \tau$, then $(F, A) \tilde{\cap} (G, B) \in \tau$.

Then, the pair $((U, E), \tau)$ is called soft multi-topological space. The members of τ are called soft multi-open sets (or τ—open soft multi-sets or simply open sets), and the conditions $[O_1]$, $[O_2]$, and $[O_3]$ are called the axioms for soft multi-open sets.

Example 8.27 Let us consider there are three universes U_1, U_2 and U_3.
   Let $U_1 = \{h_1, h_2, h_3, h_4\}$, $U_2 = \{c_1, c_2, c_3\}$, and $U_3 = \{v_1, v_2\}$. Let $\{E_{U_1}, E_{U_2}, E_{U_3}\}$ be a collection of sets of decision parameters related to the above universes, where

$$E_{U_1} = \{e_{U_1,1} = \text{expensive}, e_{U_1,2} = \text{cheap}, e_{U_1,3} = \text{wooden},$$
$$e_{U_1,4} = \text{in green surroundings}\}$$
$$E_{U_2} = \{e_{U_2,1} = \text{expensive}, e_{U_2,2} = \text{cheap}, e_{U_2,3} = \text{sporty}\},$$
$$E_{U_3} = \{e_{U_3,1} = \text{expensive}, e_{U_3,2} = \text{cheap}, e_{U_3,3} = \text{in Kuala Lumpur},$$
$$e_{U_3,4} = \text{majestic}\}.$$

Let $U = \prod_{i=1}^3 P(U_i)$, $E = \prod_{i=1}^3 E_{U_i}$, and $A^1 = \{e_1 = (e_{U_1,1}, e_{U_2,1}, e_{U_3,1}),$
$e_2 = (e_{U_1,1}, e_{U_2,2}, e_{U_3,1})\}, A^2 = \{e_1 = (e_{U_1,1}, e_{U_2,1}, e_{U_3,1}), e_3 = (e_{U_1,2}, e_{U_2,3}, e_{U_3,1})\}$
Suppose that

$$(F^1, A^1) = \{(e_1, (\{h_1, h_2\}, \{c_1, c_2\}, \{v_1\})), (e_2, (\{h_3, h_4\}, \{c_1, c_3\}, \{v_2\}))\},$$
$$(F^2, A^2) = \{(e_1, (\{h_1, h_3\}, \{c_2, c_3\}, \{v_1, v_2\})), (e_3, (\{h_2, h_4\}, \{c_1, c_2\}, \{v_2\}))\},$$
$$(F^3, A^3) = (F^1, A^1) \tilde{\cup} (F^2, A^2) = \{(e_1, (\{h_1, h_2, h_3\}, \{c_1, c_2, , c_3\}, \{v_1, v_2\})),$$
$$(e_2, (\{h_3, h_4\}, \{c_1, c_3\}, \{v_2\})), (e_3, (\{h_2, h_4\}, \{c_1, c_2\}, \{v_2\}))\},$$
$$(F^4, A^4) = (F^1, A^1) \tilde{\cap} (F^2, A^2) = \{(e_1, (\{h_1\}, \{c_2\}, \{v_1\})),$$
$$(e_2, (\{h_3, h_4\}, \{c_1, c_3\}, \{v_2\})), (e_3, (\{h_2, h_4\}, \{c_1, c_2\}, \{v_2\}))\},$$

where $A^3 = A^4 = A^1 \cup A^2 = \{e_1 = (e_{U_1,1}, e_{U_2,1}, e_{U_3,1}), e_2 = (e_{U_1,1}, e_{U_2,2}, e_{U_3,1}), e_3 = (e_{U_1,2}, e_{U_2,3}, e_{U_3,1})\}$.

Then, we observe that the subfamily $\tau_1 = \left\{ \tilde{\phi}, \tilde{E}, (F^1, A^1), (F^2, A^2), (F^3, A^3), (F^4, A^4) \right\}$ of SMS (U, E) is a soft multi-topology on (U, E), since it satisfies the necessary three axioms $[O_1]$, $[O_2]$ and $[O_3]$, and $((U, E), \tau_1)$ is a soft multi-topological space. But the subfamily $\tau_2 = \left\{ \tilde{\phi}, \tilde{E}, (F^1, A^1), (F^2, A^2) \right\}$ of SMS (U, E) is not a soft multi-topology on (U, E) since the union $(F^1, A^1) \ \tilde{\cup} \ (F^2, A^2)$ and the intersection $(F^1, A^1) \ \tilde{\cap} \ (F^2, A^2)$ do not belong to τ_2.

Definition 8.28 As every soft multi-topology on (U, E) must contain the sets $\tilde{\phi}$ and $\tilde{E}$, so the family $I = \left\{ \tilde{\phi}, \tilde{E} \right\}$ forms a soft multi-topology on (U, E). This topology is called indiscrete soft multi-set topology, and the pair $((U, E), I)$ is called an indiscrete soft multi-topological space.

Definition 8.29 Let $\mathscr{D}$ denote family of all the soft multi-subsets of (U, E). Then, we observe that $\mathscr{D}$ satisfies all the axioms for topology on (U, E). This topology is called discrete soft multi-topology, and the pair $((U, E), \mathscr{D})$ is called a discrete soft multi-topological space.

Proposition 8.30 *Let* $((U, E), \tau)$ *be a soft multi-topological space over* (U, E). *Then, the collection* $\tau_e = \{F(e) : (F, A) \in \tau\}$ *for each* $e \in E$ *defines a topology on* U.

Proof

$[O_1]$ Since $\tilde{\phi}, \tilde{E} \in \tau$ implies that $\varphi, U \in \tau_e$, for each $e \in E$.
$[O_2]$ Let $\left\{ F^k(e) : k \in K \right\} \subseteq \tau_e$, for some $\left\{ (F^k, A^k) : k \in K \right\} \subseteq \tau$. Since $\tilde{\cup}_{k \in K} (F^k, A^k) \in \tau$, $\tilde{\cup}_{k \in K} F^k(e) \in \tau_e$, for each $e \in E$.
$[O_3]$ Let $F(e), G(e) \in \tau_e$, fore some $(F, A), (G, B) \in \tau$. Since $(F, A) \ \tilde{\cap} \ (G, B) \in \tau$, $F(e) \ \tilde{\cap} \ G(e) \in \tau_e$, for each $e \in E$.

Thus, τ_e defines a topology on U for each $e \in E$. $\qquad\qquad\square$

Example 8.31 Let us consider the soft multi-topology $\tau_1 = \{\tilde{\phi}, \tilde{E}, (F^1, A^1), (F^2, A^2), (F^3, A^3), (F^4, A^4)\}$ as in the Example 8.27, where

$$F^1(e_1) = (\{h_1, h_2\}, \{c_1, c_2\}, \{v_1\}), \ F^1(e_2) = (\{h_3, h_4\}, \{c_1, c_3\}, \{v_2\})$$
$$F^2(e_1) = (\{h_1, h_3\}, \{c_2, c_3\}, \{v_1, v_2\}), \ F^2(e_3) = (\{h_2, h_4\}, \{c_1, c_2\}, \{v_2\}),$$
$$F^3(e_1) = (\{h_1, h_2, h_3\}, \{c_1, c_2, , c_3\}, \{v_1, v_2\}), \ F^3(e_2) = (\{h_3, h_4\}, \{c_1, c_3\}, \{v_2\}),$$
$$F^3(e_3) = (\{h_2, h_4\}, \{c_1, c_2\}, \{v_2\}),$$
$$F^4(e_1) = (\{h_1\}, \{c_2\}, \{v_1\}), \ F^4(e_2) = (\{h_3, h_4\}, \{c_1, c_3\}, \{v_2\}),$$
$$F^4(e_3) = (\{h_2, h_4\}, \{c_1, c_2\}, \{v_2\}).$$

It can be easily seen that $\tau_{e_1} = \{\varphi, U, (\{h_1, h_2\}, \{c_1, c_2\}, \{v_1\}), (\{h_1, h_3\}, \{c_2, c_3\},$ $\{v_1, v_2\}), (\{h_1\}, \{c_2\}, \{v_1\}), (\{h_1, h_2, h_3\}, \{c_1, c_2, , c_3\}, \{v_1, v_2\})\}, \tau_{e_2} = \{\varphi, U,$ $(\{h_3, h_4\}, \{c_1, c_3\}, \{v_2\})\}$ and $\tau_{e_3} = \{\phi, U_2, (\{h_2, h_4\}, \{c_1, c_2\}, \{v_2\})\}$ are topologies on U.

Definition 8.32 Let $((U, E), \tau_1)$ and $((U, E), \tau_2)$ be two soft multi-set topological spaces. If each $(F, A) \in \tau_1 \Rightarrow (F, A) \in \tau_2$, then τ_2 is called soft multi-set finer (stronger) topology than τ_1 and τ_1 is called soft multi-set coarser (or weaker) topology than τ_2 and denoted by $\tau_1 \tilde{\subseteq} \tau_2$.

Two soft multi-topologies, one of which is finer than other, are said to be comparable.

Example 8.33 The indiscrete soft multi-topology on (U, E) is the soft multi-coarsest (weakest), and discrete soft multi-topology on (U, E) is the soft multi-finest (strongest) of all topologies of (U, E). Any other soft multi-topology on (U, E) will be in between these two soft multi-set topologies.

Example 8.34 If we consider the topologies τ_1 as in the Example 8.27 and $\tau_3 = \{\tilde{\phi}, \tilde{E}, (F^1, A^1)\}$ on (U, E). Then, τ_1 is soft multi-finer topology than τ_3, and τ_3 is soft multi-coarser topology than τ_1, i.e. $\tau_3 \tilde{\subseteq} \tau_1$.

Theorem 8.35 *Let* $\{\tau_i : i \in I\}$ *be arbitrary collection of soft multi-topologies on* (U, E). *Then, their intersection* $\cap_{i \in I} \tau_i$ *is also a soft multi-topology on* (U, E).

Proof

$[O_1]$ Since $\tilde{\phi}, \tilde{E} \in \tau_i$, for each $i \in I$, hence $\tilde{\phi}, \tilde{E} \in \cap_{i \in I} \tau_i$

$[O_2]$ Let $\{(F^k, A^k) | k \in K\}$ be an arbitrary family of soft multi-sets, where $(F^k, A^k) \in \cap_{i \in I} \tau_i$ for each $k \in K$. Then, for each $i \in I$, $(F^k, A^k) \in \tau_i$ for $k \in K$ and for each $i \in I$, τ_i is a soft multi-topology; therefore, $\tilde{\cup}_{k \in K}(F^k, A^k) \in \tau_i$, for each $i \in I$. Hence, $\tilde{\cup}_{k \in K}(F^k, A^k) \in \cap_{i \in I} \tau_i$

$[O_3]$ Let $(F, A), (G, B) \in \cap_{i \in I} \tau_i$, and then, $(F, A), (G, B) \in \tau_i$, for each $i \in I$ and since τ_i is a soft multi-topology for each $i \in I$, therefore $(F, A) \tilde{\cap} (G, B) \in \tau_i$ for each $i \in I$. Hence $(F, A) \tilde{\cap} (G, B) \in \cap_{i \in I} \tau_i$.

Thus, $\cap_{i \in I} \tau_i$ satisfies all the axioms of topology. Hence, $\cap_{i \in I} \tau_i$ forms a topology. But union of topologies need not be a topology; we can show this with the following Example.   □

Example 8.36 The union of two soft multi-topologies may not be a soft multi-topology. If we consider the Example 4.2, then the subfamilies $\tau_3 = \{\tilde{\phi}, \tilde{E}, (F^1, A^1)\}$ and $\tau_4 = \{\tilde{\phi}, \tilde{E}, (F^2, A^2)\}$ are the soft multi-topology sets on (U, E). But their union $\tau_3 \cup \tau_4 = \{\tilde{\phi}, \tilde{E}, (F^1, A^1), (F^2, A^2)\} = \tau_2$ is not a soft multi-topology on (U, E).

Definition 8.37 Let $((U,E),\tau)$ be a soft multi-topological space over (U,E). A soft multi-subset (F,A) of (U,E) is called soft multi-closed set if its relative complement $(F,A)'$ is a member of τ.

Example 8.38 Let us consider Example 8.27, and then, the soft multi-closed sets in $((U,E),\tau_1)$ are

$$\left(\tilde{\phi}\right)' = \tilde{E}, \ (\tilde{E})' = \tilde{\phi},$$

$$\left(F^1,A^1\right)' = \{(e_1,(\{h_3,h_4\},\{c_3\},\{v_2\})),(e_2,(\{h_1,h_2\},\{c_2\},\{v_1\}))\},$$

$$\left(F^2,A^2\right)' = \{(e_1,(\{h_2,h_4\},\{c_1\},\phi)),(e_3,(\{h_1,h_3\},\{c_3\},\{v_1\}))\},$$

$$\left(F^3,A^3\right)' = \{(e_1,(\{h_4\},\phi,\phi)),(e_2,(\{h_1,h_2\},\{c_2\},\{v_1\})),$$
$$(e_3,(\{h_1,h_3\},\{c_3\},\{v_1\}))\}$$

$$\left(F^4,A^4\right)' = \{(e_1,(\{h_2,h_3,h_4\},\{c_1,c_3\},\{v_2\})),(e_2,(\{h_1,h_2\},\{c_2\},\{v_1\})),$$
$$(e_3,(\{h_1,h_2\},\{c_3\},\{v_1\}))\}.$$

Definition 8.39 Let $((U,E),\tau)$ be a soft multi-set topological space over (U,E). Then,

(1) $\tilde{\phi}$ and $\tilde{E}$ are soft multi-closed sets over (U,E).
(2) The intersection of arbitrary collection of soft multi-closed sets is a soft multi-closed set over (U,E).
(3) The union of any two soft multi-closed sets is a soft multi-closed set over (U,E).

Proof Straightforward. □

8.3 Soft Multi-Basis and Soft Multi-Sub-Basis

In this section, soft multi-basis and soft multi-sub-basis are to be introduced.

Definition 8.40 Let $((U,E),\tau)$ be a soft multi-set topological space on (U,E) and $\mathscr{B}$ be a subfamily of τ. If every element of τ can be expressed as the arbitrary soft multi-set union of some element of $\mathscr{B}$, then $\mathscr{B}$ is called soft multi-basis (in short base) for the soft multi-topology τ.

Definition 8.41 A collection $S \subseteq \tau$ is called a multi-soft sub-basis (in short sub-basis) for the topology τ if the set $\mathscr{B}(S)$ consisting of finite intersections of elements of S forms a multi-soft basis for τ.

Example 8.42 In the Example 8.27 for the topology τ_1, the subfamily $\mathscr{B} = \{\tilde{\phi}, \tilde{E}, (F^1,A^1), (F^2,A^2), (F^4,A^4)\}$ of SMS (U,E) is a multi-soft basis for the topology τ_1 and $S = \{\tilde{\phi}, \tilde{E}, (F^1, A^1), (F^2, A^2)\}$ is a multi-soft sub-basis for the topology τ_1, since $\mathcal{B}(S) = \{\tilde{\phi}, \tilde{E}, (F^1, A^1), (F^2, A^2), (F^4, A^4)\}$ is a multi-soft basis for the topology τ_1.

Theorem 8.43 *Let* $((U, E), \tau)$ *be soft multi-topological space on* (U, E). *A subfamily* $\mathcal{B}$ *of* τ *forms a base for a topology* τ *if and only if*

1. $\tilde{E} = \tilde{\cup}\{(G, B) : (G, B) \in \mathcal{B}\}$
2. *For every* $(G_1, B_1), (G_2, B_2) \in \mathcal{B}$, $(G_1, B_1) \,\tilde{\cap}\, (G_2, B_2)$ *is the union of members of* $\mathcal{B}$.

Proof **Necessity:** Let $\mathcal{B}$ be a base for a topology τ on (U, E).

(1) Since $\tilde{E} \in \tau$, we have $\tilde{E} = \tilde{\cup}\{(G, B) : (G, B) \in \mathcal{B}\}$.
(2) If $(G_1, B_1), (G_2, B_2) \in \mathcal{B}$, then $(G_1, B_1), (G_2, B_2) \in \tau$, since $\mathcal{B}$ is subfamily of τ and since τ is a topology on, therefore $(G_1, B_1) \,\tilde{\cap}\, (G_2, B_2) \in \tau$, and thus, $(G_1, B_1) \,\tilde{\cap}\, (G_2, B_2)$ is the union of members of $\mathcal{B}$.

Sufficiency: Let $\mathcal{B}$ be a family with the given properties and let τ be the family of all unions of members of $\mathcal{B}$. Now, if we can prove that τ is a topology on (U, E), then it is obvious that $\mathcal{B}$ is a base for this topology.

$[O_1]$ $\tilde{\phi}$ and $\tilde{E} \in \tau$, since $\tilde{\phi} \in \tau$ is the union of empty subcollection from $\mathcal{B}$ (i.e. $\tilde{\phi} = \tilde{\cup}\{(G, B) : (G, B) \in \tilde{\phi} \tilde{\subseteq} \mathcal{B}\}$ and $\tilde{E} \in \tau$, by condition (1). $\tilde{E} = \tilde{\cup}\{(G, B) : (G, B) \in \mathcal{B}\}$.

$[O_2]$ Let $(F^k, A^k) \in \tau$ for all k. By definition of τ, each $(F^k, A^k) = \tilde{\cup}\{(G, B) : (G, B) \in \mathcal{B}\}$, hence $\tilde{\cup}_k(F^k, A^k) = \tilde{\cup}_k(\tilde{\cup}\{(G, B) : (G, B) \in \mathcal{B}\})$ is also the union of members of $\mathcal{B}$ and so belongs to τ. Thus, τ satisfies $[O_2]$.

$[O_3]$ Let $(F^1, A^1), (F^2, A^2) \in \tau$. By definition of τ, $(F^1, A^1) = \tilde{\cup}\{(G, B) : (G, B) \in \mathcal{B}\}$ and $(F^2, A^2) = \tilde{\cup}\{(H, C) : (H, C) \in \mathcal{B}\}$, hence $(F^1, A^1) \,\tilde{\cap}\, (F^2, A^2) = (\tilde{\cup}\{(G, B) : (G, B) \in \mathcal{B}\}) \,\tilde{\cap}\, (\tilde{\cup}\{(H, C) : (H, C) \in \mathcal{B}\}) = \tilde{\cup}\{(G, B) \,\tilde{\cap}\, (H, C) : (G, B), (H, C) \in \mathcal{B}\} = \tilde{\cup}\{(G, B) \,\tilde{\cap}\, (H, C) : (G, B), (H, C) \in \mathcal{B}\}$. Condition (2) implies that $(F^1, A^1) \,\tilde{\cap}\, (F^2, A^2)$ is expressible as the union of the member of $\mathcal{B}$ and hence is a member of τ.

The topology τ obtained as above forms a base called the topology generated by the base $\mathcal{B}$ is called the topology generated by the base $\mathcal{B}$. Since the base defined as above is a subfamily of τ, i.e. members of base are open, it is called an open base. $\quad\square$

8.4 Neighbourhoods and Neighbourhood Systems

We introduce neighbourhoods and neighbourhood systems in a soft multi-topological space.

Definition 8.44 Let τ be the soft multi-topology on (U, E). A soft multi-set (F, A) in $\text{SMS}(U, E)$ is a neighbourhood of a soft multi-set (G, B) if and only if there exists an τ—open soft multi-set (H, C), i.e. $(H, C) \in \tau$ such that $(G, B) \widetilde{\subseteq} (H, C) \widetilde{\subseteq} (F, A)$.

Example 8.45 Let us consider there are three universes U_1, U_2, and U_3. Let $U_1 = \{h_1, h_2, h_3, h_4\}$, $U_2 = \{c_1, c_2, c_3\}$, and $U_3 = \{v_1, v_2\}$. Let $\{E_{U_1}, E_{U_2}, E_{U_3}\}$ be a collection of sets of decision parameters related to the above universes, where

$$E_{U_1} = \{e_{U_1,1} = \text{expensive}, e_{U_1,2} = \text{cheap}, e_{U_1,3} = \text{wooden},$$
$$e_{U_1,4} = \text{in green surroundings}\}$$
$$E_{U_2} = \{e_{U_2,1} = \text{expensive}, e_{U_2,2} = \text{cheap}, e_{U_2,3} = \text{sporty}\},$$
$$E_{U_3} = \{e_{U_3,1} = \text{expensive}, e_{U_3,2} = \text{cheap}, e_{U_3,3} = \text{in Kuala Lumpur}, e_{U_3,4} = \text{majestic}\}.$$

Let $U = \prod_{i=1}^{3} P(U_i)$, $E = \prod_{i=1}^{3} E_{U_i}$ and let

$$A = \{e_1 = (e_{U_1,1}, e_{U_2,1}, e_{U_3,1}), e_2 = (e_{U_1,1}, e_{U_2,2}, e_{U_3,1}),$$
$$e_3 = (e_{U_1,2}, e_{U_2,3}, e_{U_3,1}), e_4 = (e_{U_1,2}, e_{U_2,3}, e_{U_3,2})\}$$
$$B = \{e_1 = (e_{U_1,1}, e_{U_2,1}, e_{U_3,1}), e_2 = (e_{U_1,1}, e_{U_2,2}, e_{U_3,1})\}$$
$$C = \{e_1 = (e_{U_1,1}, e_{U_2,1}, e_{U_3,1}), e_2 = (e_{U_1,1}, e_{U_2,2}, e_{U_3,1}), e_3 - (e_{U_1,2}, e_{U_2,3}, e_{U_3,1})\}.$$

In a soft multi-topology

$$\tau = \Big\{ \widetilde{\phi}, \widetilde{E}, \{(e_1, (\{h_1, h_2\}, \{c_1, c_2\}, \{v_1\})), (e_2, (\{h_3, h_4\}, \{c_1, c_3\}, \{v_2\}))$$
$$(e_3, (\{h_1, h_2, h_3\}, \{c_2, c_3\}, \{v_2\}))\} \Big\},$$

the soft multi-set

$$(F, A) = \{(e_1, (\{h_1, h_2, h_3\}, \{c_1, c_2\}, \{v_1\})), (e_2, (\{h_2, h_3, h_4\}, \{c_1, c_3\}, \{v_1, v_2\})),$$
$$(e_3, (\{h_1, h_2, h_3\}, \{c_2, c_3\}, \{v_2\})), (e_4, (\{h_1\}, \{c_2\}, \{v_2\}))\}$$

is a neighbourhood of the soft multi-set $(G, B) = \{(e_1, (\{h_1\}, \{c_2\}, \{v_1\})), (e_2, (\{h_4\}, \{c_1, c_3\}, \{v_2\}))\}$, because there exists an τ—open soft multi-set $(H, C) = \{(e_1, (\{h_1, h_2\}, \{c_1, c_2\}, \{v_1\})), (e_2, (\{h_3, h_4\}, \{c_1, c_3\}, \{v_2\})), (e_3, (\{h_1, h_2, h_3\}, \{c_2, c_3\}, \{v_2\}))\} \in \tau$ such that $(G, B) \widetilde{\subseteq} (H, C) \widetilde{\subseteq} (F, A)$.

Theorem 8.46 *A soft multi-set (F, A) in $\text{SMS}(U, E)$ is a soft multi-open set if and only if (F, A) is a neighbourhood of each soft multi-set (G, B) contained in (F, A).*

Definition 8.47 Let $((U, E), \tau)$ be a soft multi-topological space on (U, E) and (F, A) be a soft multi-set in $\text{SMS}(U, E)$. The family of all neighbourhoods of

(F,A) is called the neighbourhood system of (F,A) up to topology and is denoted by $N_{(F,A)}$.

Theorem 8.48 *Let $((U,E),\tau)$ be a soft multi-set topological space. If $N_{(F,A)}$ is the neighbourhood system of a soft multi-set (F,A), then*

1. *$N_{(F,A)}$ is non-empty, and (F,A) is soft multi-subset of the each member of $N_{(F,A)}$.*
2. *The intersection of any two members of $N_{(F,A)}$ belongs to $N_{(F,A)}$.*
3. *Each soft multi-set which contains a member of $N_{(F,A)}$ belongs to $N_{(F,A)}$.*

Proof Straightforward. $\square$

Definition 8.49 Let $((U,E),\tau)$ be a soft multi-topological space on (U,E) and (F,A) be a soft multi-set in SMS (U,E). A collection $B_{(F,A)} \subseteq$ SMS (U,E) of subsets containing the neighbourhood of (F,A) is called a neighbourhood basis of (F,A) if

(1) Every element of $B_{(F,A)}$ is a neighbourhood of (F,A).
(2) Every neighbourhood of (F,A) contains an element of $B_{(F,A)}$ as a subset.

8.5 Interior and Closure

Here, we give the definitions of interior and closure.

Definition 8.50 Let $((U,E),\tau)$ be a soft multi-topological space on (U,E) and (F,A) be a soft multi-set in SMS(U,E). Then, the union of all soft multi-open sets contained in (F,A) is called the interior of (F,A) and is denoted by int(F,A) and defined by int$(F,A) = \tilde{\cup}\{(G,B)|(G,B)$ is a soft multi open set contained in (F, A)$\}$.

Example 8.51 Let us consider the soft multi-topology τ_1 as in the Example 8.27, and let $(F,A) = \{(e_1,(\{h_1,h_2,h_3\},\{c_1,c_2\},\{v_1\})),(e_2,(\{h_1,h_3,h_4\},\{c_1,c_3\},\{v_2\}))\}$ be a soft multi-set, and then,

$$\text{int}(F,A) = \tilde{\cup}\{(G,B)|(G,B) \text{ is a soft multi-open set contained in } (F,A)\}$$
$$= (F^1,A^1) \tilde{\cup} (F^4,A^4)$$
$$= \{(e_1,(\{h_1,h_2\},\{c_1,c_2\},\{v_1\})),(e_2,(\{h_3,h_4\},\{c_1,c_3\},\{v_2\})),$$
$$(e_3,(\{h_2,h_4\},\{c_1,c_2\},\{v_2\}))\},$$

Let (F^1,A^1) and (F^4,A^4) be the two soft multi-open sets contained in (F,A).

Theorem 8.52 *Let* $((U,E),\tau)$ *be a soft multi-set topological space on* (U,E) *and* (F,A) *be a soft multi-set in* $\text{SMS}(U,E)$. *Then,*

1. $\text{int}(F,A)$ *is an open and* $\text{int}(F,A)$ *is the largest open soft multi-set contained in* (F,A).
2. *The soft multi-set* (F,A) *is open if and only if* $(F,A) = \text{int}(F,A)$.

Proposition 8.53 *For any two soft multi-sets* (F,A) *and* (G,B) *in a soft multi-topological space* $((U,E),\tau)$ *on* (U,E),

(i) $(G,B) \tilde{\subseteq} (F,A) \Rightarrow \text{int}(G,B) \tilde{\subseteq} \text{int}(F,A)$
(ii) $\text{int}\,\tilde{\phi} = \tilde{\phi}$ and $\text{int}\,\tilde{E} = \tilde{E}$
(iii) $\text{int}(\text{int}(F,A)) = \text{int}(F,A)$
(iv) $\text{int}(F,A) \tilde{\cap} \text{int}(G,B) \tilde{\subseteq} \text{int}((F,A) \tilde{\cap} (G,B))$
(v) $\text{int}((F,A) \tilde{\cup} (G,B)) \tilde{\supseteq} \text{int}(F,A) \tilde{\cup} \text{int}(G,B)$

Proof The proof is straightforward.   □

Definition 8.54 Let $((U,E),\tau)$ be a soft multi-topological space on (U,E) and (F,A) be a soft multi-set in $\text{SMS}(U,E)$. Then, the intersection of all soft multi-closed set containing (F,A) is called the closure of (F,A) and is denoted by $\text{cl}(F,A)$ and defined by

$$\text{cl}(F,A) = \tilde{\cap}\,\{(G,B)|(G,B) \text{ is a soft multi-closed set containing } (F,A)\}.$$

Observe first that $\text{cl}(F,A)$ is a soft multi-closed set, since it is the intersection of soft multi-closed sets. Furthermore, $\text{cl}(F,A)$ is the smallest soft multi-closed set containing (F,A).

Example 8.55 Let us consider the soft multi-topology τ_1 as in the Example 8.27, and let $(F,A) = \{(e_1,(\{h_4\},\{c_3\},\{v_2\})),(e_2,(\{h_2\},\{c_2\},\{v_1\}))\}$. be a soft multi-set, and then,

$$\begin{aligned}
\text{cl}(F,A) &= \tilde{\cap}\{(G,B)|(G,B) \text{ is a soft multi-closed set containing } (F,A)\} \\
&= (F^1,A^1)' \tilde{\cap} (F^4,A^4)' \\
&= \{(e_1,(\{h_3,h_4\},\{c_3\},\{v_2\})),(e_2,(\{h_1,h_2\},\{c_2\},\{v_1\})), \\
&\quad\ (e_3,(\{h_1,h_2\},\{c_3\},\{v_1\}))\},
\end{aligned}$$

Let $(F^1,A^1)'$ and $(F^4,A^4)'$ be the two soft multi-closed sets in τ_1 containing (F,A).

Proposition 8.56 *For any two soft multi-sets* (F,A) *and* (G,B) *in a soft multi-set topological space* $((U,E),\tau)$ *on* (U,E),

(i) $\mathrm{cl}\tilde{\phi} = \tilde{\phi}$ and $\mathrm{cl}\tilde{E} = \tilde{E}$

(ii) $(F,A) \tilde{\subseteq} \mathrm{cl}(F,A)$

(iii) (F,A) is a soft multi-closed set if and only if $(F,A) = \mathrm{cl}(F,A)$

(iv) $\mathrm{cl}(\mathrm{cl}\,(F,A)) = \mathrm{cl}\,(F,A)$

(v) $(G,B) \tilde{\subseteq} (F,A) \Rightarrow \mathrm{cl}(G,B) \tilde{\subseteq} \mathrm{cl}\,(F,A)$

(vi) $\mathrm{cl}\,((F,A) \tilde{\cap} (G,B)) \tilde{\subseteq} \mathrm{cl}(F,A) \tilde{\cap} \mathrm{cl}\,(G,B)$

(vii) $\mathrm{cl}((F,A) \tilde{\cup} (G,B)) = \mathrm{cl}\,(F,A) \tilde{\cup} \mathrm{cl}(G,B)$

Proof The proof is straightforward.   □

Theorem 8.57 *Let* $((U,E),\tau)$ *be a soft multi-set topological space on* (U,E), *and let* (F,A) *be a soft multi-set in* $\mathrm{SMS}(U,E)$. *Then,*

(i) $(\mathrm{cl}(F,A))' = \mathrm{int}((F,A)')$

(ii) $(\mathrm{int}(F,A))' = \mathrm{cl}((F,A)')$

Proof Straightforward.   □

Definition 8.58 Let $((U,E),\tau)$ be a soft multi-topological space on (U,E) and (G,B) be a soft multi-set in $\mathrm{SMS}(U,E)$. The soft multi-point $e_{(F,A)}$ in $\mathrm{SMS}(U,E)$ is called a soft multi-interior point of a soft multi-set (G,B) if there exists a soft multi-open set (H,C), such that $e_{(F,A)} \tilde{\in} (H,C) \tilde{\subseteq} (G,B)$.

Example 8.59 Let us consider the soft multi-topology τ_1 as in the Example 8.27 and let $(F,A) = \{(e_1,(\{h_1,h_2,h_3\},\{c_1,c_2\},\{v_1\})),(e_2,(\{h_1,h_3,h_4\},\{c_1,c_3\},\{v_2\}))\}$ be a soft multi-set, and then, $e_{1_{(F,A)}} = \{(e_1,(\{h_1\},\{c_1\},\{v_1\}))\}$ is a soft multi-interior point of the soft multi-set (F,A), since there exists a soft multi-open set $(F^1,A^1) \in \tau_1$, such that $e_{(F,A)} \tilde{\in} (F^1,A^1) \tilde{\subseteq} (F,A)$. But $e_{2_{(F,A)}} = \{(e_2,(\{h_1,h_3,h_4\}, \phi,\phi))\}$ is not a soft multi-interior point of the soft multi-set (F,A), since there does not exist a soft multi-open set (H,C), such that $e_{2_{(F,A)}} \tilde{\in} (H,C) \tilde{\subseteq} (F,A)$.

Proposition 8.60 *Let* $((U,E),\tau)$ *be a soft multi-topological space on* (U,E) *and* (G,B) *be a soft multi-open set in* $\mathrm{SMS}(U,E)$. *Then, every soft multi-point* $e_{(F,A)} \tilde{\in} (G,B)$ *is a soft multi-interior point.*

Proof The proof is straightforward.   □

Definition 8.61 Let $((U,E),\tau)$ be a soft multi-topological space on (U,E) and (F,A) be a soft multi-set in $\mathrm{SMS}(U,E)$. Then, we defined a soft multi-set associate with (F,A) over (U,E) denoted by $(\mathrm{cl}(F),A)$ and defined by $\mathrm{cl}(F)(e) = \mathrm{cl}(F(e))$, where $\mathrm{cl}(F(e))$ is the closer of $F(e)$ in τ_e for each $e \in A$.

Proposition 8.62 *Let* $((U,E),\tau)$ *be a soft multi-topological space on* (U,E) *and* (F,A) *be a soft multi-set in* $\mathrm{SMS}(U,E)$. *Then,* $(\mathrm{cl}(F),A) \tilde{\subseteq} \mathrm{cl}(F,A)$.

Proof For any $e \in A$, $\mathrm{cl}(F(e))$ is the smallest closed set in (U,τ_e), which contains $F(e)$. Moreover, if $\mathrm{cl}(F,A) = (G,B)$, then $G(e)$ is also closed set in (U,τ_e)

containing $F(e)$. This implies that $\mathrm{cl}(F)(e) = \mathrm{cl}(F(e)) \subseteq G(e)$. Thus, $(\mathrm{cl}(F), A) \,\tilde{\subseteq}\, \mathrm{cl}(F, A)$. $\qquad\square$

Corollary 8.63 *Let* $((U, E), \tau)$ *be a soft multi-topological space on* (U, E) *and* (F, A) *be a soft multi-set in* $\mathrm{SMS}(U, E)$. *Then,* $(\mathrm{cl}(F), A) = \mathrm{cl}(F, A)$ *if and only if* $(\mathrm{cl}(F), A)' \in \tau$.

Proof If $(\mathrm{cl}(F), A) = \mathrm{cl}(F, A)$, then $(\mathrm{cl}(F), A) = \mathrm{cl}(F, A)$ is a soft multi-closed set and so $(\mathrm{cl}(F), A)' \in \tau$.

Conversely, if $(\mathrm{cl}(F), A)' \in \tau$, then $(\mathrm{cl}(F), A)$ is a soft multi-closed set containing (F, A). By Proposition 7.13 $(\mathrm{cl}(F), A) \,\tilde{\subseteq}\, \mathrm{cl}(F, A)$ and by the definition of soft multi-closure of (F, A), any soft multi-closed set over which contains (F, A) will contain $\mathrm{cl}(F, A)$. This implies that $\mathrm{cl}(F, A) \,\tilde{\subseteq}\, (\mathrm{cl}(F), A)$. Thus, $(\mathrm{cl}(F), A) = \mathrm{cl}(F, A)$. $\qquad\square$

8.6 Soft Multi-Subspace Topology

In this section, we introduce soft multi-subspace topology.

Theorem 8.64 *Let* $((U, E), \tau)$ *be a soft multi-topological space on* (U, E) *and* (F, A) *be a soft multi-set in* $\mathrm{SMS}(U, E)$. *Then, the collection* $\tau_{(F,A)} = \{(F, A) \,\tilde{\cap}\, (G, B) | (G, B) \in \tau\}$ *is a soft multi-topology on the soft multi-set* (F, A).

Proof

$[O_1]$ Since $\tilde{\phi}, \tilde{E} \in \tau$, therefore $(F, A) = (F, A) \,\tilde{\cap}\, \tilde{E}$ and $\tilde{\phi}_{(F,A)} = (F, A) \,\tilde{\cap}\, \tilde{\phi}$ and therefore $\tilde{\phi}_{(F,A)}, (F, A) \in \tau_{(F,A)}$.

$[O_2]$ Let $\{(F^k, A^k) | k \in K\}$ be an arbitrary family of soft multi-open sets in $\tau_{(F,A)}$, and then, for each $k \in K$, there exist $(G^k, B^k) \in \tau$ such that $(F^k, A^k) = (F, A) \,\tilde{\cap}\, (G^k, B^k)$.

Now, $\tilde{\cup}_{k \in K}(F^k, A^k) = \tilde{\cup}_{k \in K}((F, A) \,\tilde{\cap}\, (G^k, B^k)) = (F, A) \,\tilde{\cap}\, (\tilde{\cup}_{k \in K}(G^k, B^k))$ and since $\tilde{\cup}_{k \in K}(G^k, B^k) \in \tau \Rightarrow \tilde{\cup}_{k \in K}(F^k, A^k) \in \tau_{(F,A)}$.

$[O_3]$ Let (F^1, A^1) and (F^2, A^2) are the two soft multi-open sets in $\tau_{(F,A)}$, and then for each $i = 1, 2$, there exist $(G^i, B^i) \in \tau$ such that $(F^i, A^i) = (F, A) \,\tilde{\cap}\, (G^i, B^i)$. Now, $(F^1, A^1) \,\tilde{\cap}\, (F^2, A^2) = ((F, A) \,\tilde{\cap}\, (G^1, B^1)) \,\tilde{\cap}\, ((F, A) \,\tilde{\cap}\, (G^2, B^2)) = (F, A) \,\tilde{\cap}\, ((G^1, B^1) \,\tilde{\cap}\, (G^2, B^2))$, and since $(G^1, B^1) \,\tilde{\cap}\, (G^2, B^2) \in \tau$, thus $(F^1, A^1) \,\tilde{\cap}\, (F^2, A^2) \in \tau_{(F,A)}$.

Thus, $\tau_{(F,A)}$ is a soft multi-topology on the soft multi-set (F, A). $\qquad\square$

Definition 8.65 Let $((U, E), \tau)$ be an soft multi-topological space on (U, E) and (F, A) be an soft multi-set in $\mathrm{SMS}(U, E)$. Then, the soft multi-topology

$\tau_{(F,A)} = \{(F,A) \,\tilde{\cap}\, (G,B) | (G,B) \in \tau\}$ is called soft multi-subspace topology, and $((F,A), \tau_{(F,A)})$ is called soft multi-topological subspace of $((U,E), \tau)$.

Example 8.66 Let us consider the soft multi-topological space $((U,E), \tau_1)$ given in the Example 8.27 and let a soft multi-set be $(F,A) = \{(e_1, (\{h_1, h_4\}, \{c_1, c_3\}, \{v_1\})), (e_4, (\{h_2, h_3, h_4\}, \{c_1, c_3\}, \{v_1, v_2\}))\}$, where $A = \{e_1 = (e_{U_1,1}, e_{U_2,1}, e_{U_3,1}), e_4 = (e_{U_1,1}, e_{U_2,2}, e_{U_3,3})\}$, then

$$
\begin{aligned}
(G^1, B^1) &= (F,A) \,\tilde{\cap}\, (F^1, A^1) \\
&= \{(e_1, (\{h_1\}, \{c_1\}, \{v_1\})), (e_2, (\{h_3, h_4\}, \{c_1, c_3\}, \{v_2\})), \\
&\quad (e_4, (\{h_2, h_3, h_4\}, \{c_1, c_3\}, \{v_1, v_2\}))\} \\
(G^2, B^2) &= (F,A) \,\tilde{\cap}\, (F^2, A^2) \\
&= \{(e_1, (\{h_1\}, \{c_3\}, \{v_1\})), (e_3, (\{h_2, h_4\}, \{c_1, c_2\}, \{v_2\})), \\
&\quad (e_4, (\{h_2, h_3, h_4\}, \{c_1, c_3\}, \{v_1, v_2\}))\} \\
(G^3, B^3) &= (F,A) \,\tilde{\cap}\, (F^3, A^3) \\
&= \{(e_1, (\{h_1\}, \{c_1, c_3\}, \{v_1\})), (e_2, (\{h_3, h_4\}, \{c_1, c_3\}, \{v_2\})), \\
&\quad (e_3, (\{h_2, h_4\}, \{c_1, c_2\}, \{v_2\})), (e_4, (\{h_2, h_3, h_4\}, \{c_1, c_3\}, \{v_1, v_2\}))\} \\
(G^4, B^4) &= (F,A) \,\tilde{\cap}\, (F^4, A^4) \\
&= \{(e_1, (\{h_1\}, \phi, \{v_1\})), (e_2, (\{h_3, h_4\}, \{c_1, c_3\}, \{v_2\})), \\
&\quad (e_3, (\{h_2, h_4\}, \{c_1, c_2\}, \{v_2\})), (e_4, (\{h_2, h_3, h_4\}, \{c_1, c_3\}, \{v_1, v_2\}))\}.
\end{aligned}
$$

Then, $\tau_{1_{(F,A)}} = \left\{ \tilde{\phi}_{(F,A)}, (F,A), (G^1, B^1), (G^2, B^2), (G^3, B^3), (G^4, B^4) \right\}$, where $B^1 = A \cup A^1 = \{e_1 = (e_{U_1,1}, e_{U_2,1}, e_{U_3,1}), e_2 = (e_{U_1,1}, e_{U_2,2}, e_{U_3,1}), e_4 = (e_{U_1,1}, e_{U_2,2}, e_{U_3,3})\}$, $B^2 = A \cup A^2 = \{e_1 = (e_{U_1,1}, e_{U_2,1}, e_{U_3,1}), \quad e_3 = (e_{U_1,2}, e_{U_2,3}, e_{U_3,1}), \cdot \quad e_4 = (e_{U_1,1}, e_{U_2,2}, e_{U_3,3})\}$, $B^3 = A \cup A^3 = B^4 = A \cup A^4 = \{e_1 = (e_{U_1,1}, e_{U_2,1}, e_{U_3,1}), e_2 = (e_{U_1,1}, e_{U_2,2}, e_{U_3,1}), \cdot \quad e_3 = (e_{U_1,2}, e_{U_2,3}, e_{U_3,1}), e_4 = (e_{U_1,1}, e_{U_2,2}, e_{U_3,3})\}$ is a soft multi-subspace topology for τ_1, and $((F,A), \tau_{1_{(F,A)}})$ is called a soft multi-subspace of $((U,E), \tau_1)$.

Theorem 8.67 *Let $((U,E), \tau)$ be a soft multi-topological space on (U,E), and let $\mathscr{B}$ be a soft multi-basis for τ and (F,A) be a soft multi-set in $\mathrm{SMS}(U,E)$. Then, the family $\mathscr{B}_{(F,A)} = \{(F,A) \,\tilde{\cap}\, (G,B) | (G,B) \in \mathscr{B}\}$ is a soft multi-basis for soft multi-subspace topology $\tau_{(F,A)}$.*

Proof Let $(H,D) \in \tau_{(F,A)}$, and then, there exists a soft multi-set $(G,B) \in \tau$, such that $(H,D) = (F,A) \,\tilde{\cap}\, (G,B)$. Since $\mathscr{B}$ is a base for τ, there exists subcollection $\{(G^i, B^i) | i \in I\}$ of $\mathscr{B}$ such that $(G,B) = \tilde{\cup}_{i \in I}(G^i, B^i)$. Therefore, $(H,D) = (F,A) \,\tilde{\cap}\, (G,B) = (F,A) \,\tilde{\cap}\, (\tilde{\cup}_{i \in I}(G^i, B^i)) = \tilde{\cup}_{i \in I}((F,A) \,\tilde{\cap}\, (G^i, B^i))$. Since $(F,A) \,\tilde{\cap}\, (G^i, B^i) \in \mathscr{B}_{(F,A)}$, which implies that $\mathscr{B}_{(F,A)}$ is a soft multi-basis for the soft multi-subspace topology $\tau_{(F,A)}$. $\qquad\square$

Theorem 8.68 Let $((U,E),\tau)$ be a soft multi-topological space on (U,E) and $((F,A),\tau^*)$ be a soft multi-topological subspace of $((U,E),\tau)$ and $((G,B),\tau^{**})$ be a soft multi-topological subspace of $((F,A),\tau^*)$. Then, $((G,B),\tau^{**})$ is also a soft multi-topological subspace of $((U,E),\tau)$.

Proof The proof is straightforward. $\qquad\qquad\qquad\qquad\qquad\qquad\qquad\qquad\square$

8.7 Soft Multi-Compact Spaces

In this section, we define soft multi-cover and soft multi-compact space.

Definition 8.69 Let $((U,E),\tau)$ be a soft multi-topological space on (U,E), and let (F,A) be any soft multi-set in $\mathrm{SMS}(U,E)$. Then, a subfamily Ω of $\mathrm{SMS}(U,E)$ is called a soft multi-cover for (F,A) if and only if $(F,A) \tilde{\subseteq} \tilde{\cup}\{(G,B)|(G,B) \in \Omega\}$, and we say that Ω covers (F,A).

Definition 8.70 If a subcollection of soft multi-cover Ω also covers (F,A), then it is called an soft multi-subcover of Ω for (F,A).

Definition 8.71 If the members of soft multi-cover Ω are open, then Ω is called soft multi-open cover.

Definition 8.72 If the members of soft multi-cover Ω are finite in number, then it is called finite soft multi-cover.

Definition 8.73 Let $((U,E),\tau)$ be a soft multi-topological space on (U,E). A soft multi-set (F,A) in $\mathrm{SMS}(U,E)$ is called soft multi-compact set if and only if every **soft multi**-open covers of (F,A) have a finite soft multi-subcover.

Definition 8.74 A soft multi-topological space $((U,E),\tau)$ is called soft multi-compact space if and only if $\tilde{E}$ is soft multi-compact space.

Definition 8.75 Let $((U,E),\tau)$ be a soft multi-topological space on (U,E). A subfamily Ω of $\mathrm{SMS}(U,E)$ has the finite intersection property if and only if the interaction of any finite subcollection of Ω is not null soft multi-set.

Theorem 8.76 *A soft multi-topological space $((U,E),\tau)$ is soft multi-compact space if and only if for every family of soft multi-closed subsets with finite intersection property has non-null intersection.*

Proof Let $((U,E),\tau)$ be a soft multi-compact space and let $\{(F^k,A^k)|k \in K\}$ be an arbitrary family of soft multi-closed sets in τ with finite intersection property.

If possible let $\tilde{\cap}_{k\in K}(F^k,A^k) = \tilde{\phi}$, then by taking complements, $(\tilde{\cap}_{k\in K}(F^k,A^k))^c$ $= (\tilde{\phi})^c$, i.e. $\tilde{\cup}_{k\in K}(F^k,A^k)^c = \tilde{E}$. So that $\{(F^k,A^k)^c|k \in K\}$ forms a soft multi-open cover for $\tilde{E}$. Since $\tilde{E}$ is compact, there is a finite **soft multi**-subcover $\{(F^i,A^i)^c|i = 1,2,3,\ldots,n\}$, such that $\tilde{E} = \tilde{\cup}_{i=1}^{n}(F^i,A^i)^c$. Then, by taking complements, $(\tilde{E})^c = (\tilde{\cup}_{i=1}^n (F^i, A^i)^c)^c$, i.e. $\tilde{\phi} = \tilde{\cap}_{i=1}^n (F^i, A^i)$. Thus, $\{(F^k, A^k) | k \in K\}$ does not have the finite intersection property, which is contrary to our assumption. Hence, $\tilde{\cap}_{k \in K} (F^k, A^k) \neq \tilde{\phi}$.

Conversely, let every family of soft multi-closed subsets in $((U, E), \tau)$ with finite intersection property has non-null intersection in $((U, E), \tau)$. Now, suppose that $((U, E), \tau)$ is not soft multi-compact space. Then, there is a soft multi-open cover $\{(G^k, B^k) | k \in K\}$ of $\tilde{E}$ that has no finite soft multi-subcover, i.e. $\tilde{E} \neq \tilde{\cup}_{i=1}^n (G^i, B^i)$, then by taking complements, $(\tilde{E})^c \neq (\tilde{\cup}_{i=1}^n (G^i, B^i))^c$, i.e. $\tilde{\phi} \neq \tilde{\cap}_{i=1}^n (G^i, B^i)^c$, which implies $\{(G^k, B^k)^c | k \in K\}$ has the finite intersection property. But by soft multi-cover property $\tilde{E} = \tilde{\cup}_{k \in K} (G^k, B^k)$, then by taking complements, $\tilde{\cap}_{k \in K} (G^k, B^k)^c = \tilde{\phi}$, i.e. the intersection of all members of the family of soft multi-closed sets is null soft multi-set, which contradicting the given condition. Hence, $((U, E), \tau)$ is soft multi-compact space. $\qquad \square$

Theorem 8.77 *Let* $((U, E), \tau)$ *be a soft multi-compact space, and let* (F, A) *be a soft multi-closed sets in* τ. *Then, the soft multi-closed subspace* $((F, A), \tau_{(F,A)})$ *of* $((U, E), \tau)$ *is soft multi-compact space.*

Proof Let $((U, E), \tau)$ be a soft multi-compact space, and let (F, A) be a soft multi-closed sets in τ. Let $\{(F^k, A^k) | k \in K\}$ be an arbitrary family of soft multi-closed sets in $((F, A), \tau_{(F,A)})$ with finite intersection property. Then, soft multi-sets (F^k, A^k) for each $k \in K$ are soft multi-closed sets in $((U, E), \tau)$, since (F, A) is a soft multi-closed sets in τ. Thus, $\{(F^k, A^k) | k \in K\}$ is a family of soft multi-closed sets in $((U, E), \tau)$, possessing finite intersection property, and as $((U, E), \tau)$ is soft multi-set compact, it follows that $\tilde{\cap}_{k \in K} (F^k, A^k) \neq \tilde{\phi}$ (by Theorem 9.8). This implies that the soft multi-closed subspace $((F, A), \tau_{(F,A)})$ of $((U, E), \tau)$ is soft multi-compact space. $\qquad \square$

References

1. Alkhazaleh, S., Salleh, A.R., Hassan, N.: Soft multisets theory. Appl. Math. Sci. **5**(72), 3561–3573 (2011)
2. Alkhazaleh, S., Salleh, A.R.: Fuzzy soft multisets theory. Abstr. Appl. Anal. **2012**, 20 p, Article ID 350603. doi:10.1155/2012/350603 (Hindawi Publishing Corporation)
3. Babitha, K.V., John, S.J.: On soft multi sets. Ann. Fuzzy Math. Inform. Accepted 5 May 2012
4. Ali, M.I., Feng, F., Liu, X., Minc, W.K., Shabir, M.: On some new operations in soft set theory. Comput. Math Appl. **57**, 1547–1553 (2009)
5. Maji, P.K., Roy, A.R., Biswas, R.: Fuzzy soft sets. J. Fuzzy Math. **9**(3), 589–602 (2001)
6. Maji, P.K., Roy, A.R., Biswas, R.: Soft set theory. Comput. Math Appl. **45**(4–5), 555–562 (2003)
7. Molodtsov, D.: Soft set theory-first results. Comput. Math Appl. **37**(4–5), 19–31 (1999)

8. Mukherjee, A., Saha, A., Das, A.: Interval valued intuitionistic fuzzy soft multisets and their relations, Ann. Fuzzy Math. Inform. Accepted in April 2013
9. Mukherjee, A., Das, A., Saha, A.: Soft multi topological spaces. Ann. Fuzzy Math. Inform. (Submitted)

Chapter 9
Soft Interval-Valued Intuitionistic Fuzzy Rough Sets

Abstract The vagueness or the representation of imperfect knowledge has been a problem for a long time for the mathematicians. There are many mathematical tools for dealing with uncertainties; some of them are fuzzy set theory, rough set theory, and soft set theory. In this chapter, the concept of soft interval-valued intuitionistic fuzzy rough sets is introduced. Also some properties based on soft interval-valued intuitionistic fuzzy rough sets are presented. Also a soft interval-valued intuitionistic fuzzy rough set-based multi-criteria group decision-making scheme is presented. The proposed scheme is illustrated by an example regarding the car selection problem.

Keywords Soft set · Rough set · Soft rough set · Soft fuzzy rough set · Soft interval-valued intuitionistic fuzzy rough set · Decision-making

The soft set theory, initiated by Molodtsov [18] in 1999, is a completely generic mathematical tool for modelling vague concepts. In soft set theory, there is no limited condition to the description of objects; so, researchers can choose the form of parameters they need, which greatly simplifies the decision-making process and make the process more efficient in the absence of partial information. Although many mathematical tools are available for modelling uncertainties such as probability theory, fuzzy set theory, rough set theory, interval-valued mathematics, etc., there are inherent difficulties associated with each of these techniques. Moreover, all these techniques lack in the parameterisation of the tools and hence, they could not be applied successfully in tackling problems especially in areas such as economic, environmental, and social problems domains. Soft set theory is standing in a unique way in the sense that it is free from the above difficulties and it has a unique scope for many applications in a multidimensional way.

Soft set theory has a rich potential for application in many directions, some of which are reported by Molodtsov [18] in his work. He successfully applied soft set theory in areas such as the smoothness of functions, game theory, operations research, and Riemann integration. Later on, Maji et al. [14] presented some new

© Springer India 2015
A. Mukherjee, *Generalized Rough Sets*, Studies in Fuzziness
and Soft Computing 324, DOI 10.1007/978-81-322-2458-7_9

definitions on soft sets such as subset, union, intersection, and complements of soft sets and discussed in detail the application of soft set in decision-making problem. Based on the analysis of several operations on soft sets introduced in [14], Ali et al. [2] presented some new algebraic operations for soft sets and proved that certain De Morgan's law holds in soft set theory with respect to these new definitions. Chen et al. [5] presented a new definition of soft set parameterisation reduction and compared this definition with the related concept of knowledge reduction in the rough set theory. Kong et al. [13] introduced the definition of normal parameter reduction into soft sets and then presented a heuristic algorithm to compute normal parameter reduction of soft sets. By amalgamating the soft sets and algebra, Aktas and Cagman [1] introduced the basic properties of soft sets, compared soft sets to the related concepts of fuzzy sets [26] and rough sets [19], and pointed out that every fuzzy set and every rough set may be considered as a soft set. Jun [12] applied soft sets to the theory of BCK/BCI-algebra and introduced the concept of soft BCK/BCI-algebra. Feng et al. [7] defined soft semi-rings and several related notions to establish a connection between the soft sets and semi-rings. Sun et al. [21] presented the definition of soft modules and constructed some basic properties using modules and Molodtsov's definition of soft sets. Maji et al. [15] presented the concept of the fuzzy soft set which is based on a combination of the fuzzy set and soft set models. Roy and Maji [20] presented a fuzzy soft set theoretic approach towards a decision-making problem. Yang et al. [22] defined the operations on fuzzy soft sets, which are based on three fuzzy logic operations: negation, triangular norm, and triangular conorm. Xiao et al. [24] proposed a combined forecasting approach based on fuzzy soft set theory. Yang et al. [23] introduced the concept of interval-valued fuzzy soft set, and a decision-making problem was analysed by the interval-valued fuzzy soft set. Feng et al. [8] presented an adjustable approach to fuzzy soft set-based decision-making and give some illustrative examples. The notion of intuitionistic fuzzy set was initiated by Atanassov [3] as a generalisation of fuzzy set. Combining soft sets with intuitionistic fuzzy sets, Maji et al. [16] introduced intuitionistic fuzzy soft sets, which are rich potentials for solving decision-making problems. The notion of the interval-valued intuitionistic fuzzy set was introduced by Atanassov and Gargov [4]. In 2010, Jiang et al. [11] introduced the concept of interval-valued intuitionistic fuzzy soft sets.

Over the years, the theories of fuzzy sets and rough sets have become much closer to each other for practical needs to use both of these two theories complementarily for managing uncertainty that arises from inexact, noisy, or incomplete information. Hybrid models combining fuzzy set with rough sets have arisen in various guises in different settings. For instance, based on the equivalence relation on the universe of discourse, Dubois and Prade [6] introduced the lower and upper approximation of fuzzy sets in a Pawlak's approximation space [19] and obtained a new notion called rough fuzzy sets. Alternatively, a fuzzy similarity relation can be used to replace an equivalence relation, and the resulting notion is called fuzzy rough sets [6]. In general, a rough fuzzy set is the approximation of a fuzzy set in a crisp approximation space, whereas a fuzzy rough set is the approximation of a crisp set or fuzzy set in a fuzzy approximation space. Feng et al. [9] provided a framework to combine rough sets and soft sets all together, which gives rise to several interesting new concepts such as soft rough sets and rough soft sets. A rough soft set is the approximation of a soft set in a Pawlak's approximation space, where as a soft rough set is based on soft rough approximations in a soft approximation space. Feng [10] presented a soft rough set-based multi-criteria group decision-making scheme. Motivated by Dubois and Prade's original idea about rough fuzzy set, Feng et al. [9] introduced lower and upper soft rough approximations of fuzzy sets in a soft approximation space and obtained a new hybrid model called soft rough fuzzy set. By employing a fuzzy soft set to granulate the universe of discourse, Meng et al. [17] introduced a more general model called soft fuzzy rough set.

The aim of this chapter was to introduce the concept of soft interval-valued intuitionistic fuzzy rough sets. Also some properties based on soft interval-valued intuitionistic fuzzy rough sets are presented. Finally, a soft interval-valued intuitionistic fuzzy rough set-based multi-criteria group decision-making scheme is presented. The proposed scheme is illustrated by an example regarding the car selection problem.

9.1 Preliminaries

This section presents a review of some fundamental notions of fuzzy sets, soft sets, and their combinations and generalisations. We refer to [1–4, 14–16, 18, 26] for details.

The theory of fuzzy sets initiated by Zadeh provides an appropriate framework for representing and processing vague concepts by allowing partial memberships. Since the establishment, this theory has been actively studied by both mathematicians and computer scientists. Many applications of fuzzy set theory have arisen over the years, for instance, fuzzy logic, fuzzy neural networks, fuzzy automata, fuzzy control systems, and so on.

Definition 9.1 [26] Let X denote a non-empty set. Then, a *fuzzy set* A on X is a set having the form

$$A = \{(x, \mu_A(x)) : x \in X\}$$

where the function $\mu_A \colon X \to [0, 1]$ is called the membership function and $\mu_A(x)$ represents the degree of membership of each element $x \in X$.

Definition 9.2 [18] Let U be the universe set and E be a set of parameters. Let $P(U)$ denotes the power set of U and $A \subseteq E$. Then, the pair (F, A) is called a *soft set* over U, where F is a mapping given by $F \colon A \to P(U)$.

In other words, the soft set is not a kind of set, but a parameterised family of subsets of U. For $e \in A$, $F(e) \subseteq U$ may be considered as the set of e-approximate elements of the soft set (F, A).

Maji et al. [15] initiated the study on hybrid structures involving both fuzzy sets and soft sets. They introduced the notion of fuzzy soft sets, which can be seen as a fuzzy generalisation of (crisp) soft sets.

Definition 9.3 [15] Let U be the universe set and E be a set of parameters. Let I^U be the set of all fuzzy subsets of U and $A \subseteq E$. Then, the pair (F, A) is called a *fuzzy soft set* over U, where F is a mapping given by $F: A \to I^U$.

It is easy to see that every (classical) soft set may be considered as a fuzzy soft set [29]. For $e \in A$, $F(e)$ is a fuzzy subset of U and is called the fuzzy value set of the parameter e. Let us denote $\mu_{F(e)}(x)$ by the membership degree that object x holds parameter e, where $e \in A$ and $x \in U$. Then, $F(e)$ can be written as a fuzzy set such that $F(e) = \{(x, \mu_{F(e)}(x)): x \in U\}$.

Before introduce the notion of the intuitionistic fuzzy soft sets, let us give the concept of intuitionistic fuzzy set [3].

Definition 9.4 [3] Let X be a non-empty set. Then, an *intuitionistic fuzzy set (IFS for short)* A is an object having the form

$$A = \{(x, \mu_A(x), \gamma_A(x)): x \in X\}$$

where the functions $\mu_A: X \to [0, 1]$ and $\gamma_A: X \to [0, 1]$ represent the degree of membership and the degree of non-membership, respectively, of each element $x \in X$ and $0 \le \mu_A(x) + \gamma_A(x) \le 1$ for each $x \in X$.

By introducing the concept of intuitionistic fuzzy sets into the theory of soft sets, Maji et al. [16] proposed the concept of the intuitionistic fuzzy soft sets as follows:

Definition 9.5 [16] Let U be the universe set and E be a set of parameters. Let IF^U be the set of all intuitionistic fuzzy subsets of U and $A \subseteq E$. Then, the pair (F, A) is called an *intuitionistic fuzzy soft set* over U, where F is a mapping given by $F: A \to \mathrm{IF}^U$.

For $e \in A$, $F(e)$ is an intuitionistic fuzzy subset of U and is called the intuitionistic fuzzy value set of the parameter e. Let us denote $\mu_{F(e)}(x)$ by the membership degree that object x holds parameter e and $\gamma_{F(e)}(x)$ by the membership degree that object x does not hold parameter e, where $e \in A$ and $x \in U$. Then, $F(e)$ can be written as an intuitionistic fuzzy set such that $F(e) = \{(x, \mu_{F(e)}(x), \gamma_{F(e)}(x)): x \in U\}$. If $\forall x \in U$, $\gamma_{F(e)}(x) = 1 - \mu_{F(e)}(x)$, then $F(e)$ will generated to be a standard fuzzy set and then (F, A) will be generated to be a traditional fuzzy soft set.

Now, before introduce the notion of the interval-valued intuitionistic fuzzy soft sets, let us give the concept of interval-valued intuitionistic fuzzy set which was first introduced by Atanassov and Gargov [4]. Actually an interval-valued intuitionistic fuzzy set is characterised by an interval-valued membership degree and an interval-valued non-membership degree.

Definition 9.6 [4] An *interval-valued intuitionistic fuzzy set (IVIFS for short)* A on an universe set U is defined as the object of the form $A = \{\langle x, \mu_A(x), \gamma_A(x) \rangle: x \in U\}$, where $\mu_A: U \rightarrow \text{Int}([0, 1])$ and $\gamma_A: U \rightarrow \text{Int}([0, 1])$ are the functions such that the condition: $\forall x \in U$, $\sup \mu_A(x) + \sup \gamma_A(x) \leq 1$ is satisfied (where $\text{Int}[0, 1]$ is the set of all closed intervals of $[0, 1]$).

We denote the class of all interval-valued intuitionistic fuzzy sets on U by IVIFS^U.

The union and intersection of the interval-valued intuitionistic fuzzy sets are defined as follows:

Let $A, B \in \text{IVIFS}^U$. Then

- the *union* of A and B is denoted by $A \cup B$, where

$$A \cup B = \{(x, [\max(\inf \mu_A(x), \inf \mu_B(x)), \max(\sup \mu_A(x), \sup \mu_B(x))],$$
$$[\min(\inf \gamma_A(x), \inf \gamma_B(x)), \min(\sup \gamma_A(x), \sup \gamma_B(x))]) : x \in U\}$$

- the *intersection* of A and B is denoted by $A \cap B$, where

$$A \cap B = \{(x, [\min(\inf \mu_A(x), \inf \mu_B(x)), \min(\sup \mu_A(x), \sup \mu_B(x))],$$
$$[\max(\inf \gamma_A(x), \inf \gamma_B(x)), \max(\sup \gamma_A(x), \sup \gamma_B(x))]) : x \in U\}$$

Atanassov and Gargov show in [4] that $A \cup B$ and $A \cap B$ are again IVIFSs.

Definition 9.7 [11] Let U be the universe set and E be a set of parameters. Let IVIFS^U be the set of all interval-valued intuitionistic fuzzy sets on U and $A \subseteq E$. Then, the pair (F, A) is called an *interval-valued intuitionistic fuzzy soft set* *(IVIFSS for short)* over U, where F is a mapping given by $F: A \rightarrow \text{IVIFS}^U$.

In other words, an interval-valued intuitionistic fuzzy soft set is a parameterised family of interval-valued intuitionistic fuzzy subsets of U. For any parameter $e \in A$, $F(e)$ can be written as an interval-valued intuitionistic fuzzy set such that $F(e) = \{(x, \mu_{F(e)}(x), \gamma_{F(e)}(x)): x \in U\}$, where $\mu_{F(e)}(x)$ is the interval-valued fuzzy membership degree that object x holds parameter e and $\gamma_{F(e)}(x)$ is the interval-valued fuzzy membership degree that object x does not hold parameter e.

9.2 Rough Sets, Rough Fuzzy Sets, Soft Rough Sets, and Rough Soft Sets

The rough set theory provides a systematic method for dealing with vague concepts caused by indiscernability in situation with incomplete information or a lack of knowledge. The rough set philosophy is founded on the assumption that with every object in the universe, we associate some information (data\knowledge). In general, a fuzzy set may be viewed as a class with unsharp boundaries, whereas a rough set is a coarsely described crisp set [25].

Definition 9.8 [19] Let R be an equivalence relation on the universal set U. Then, the pair (U, R) is called a Pawlak's approximation space. An equivalence class of R containing x will be denoted by $[x]_R$. Now for $X \subseteq U$, the lower and upper approximations of X with respect to (U, R) are denoted by, respectively, R_*X and R^*X and are defined by

$$R_*X = \{x \in U : [x]_R \subseteq X\},$$
$$R^*X = \{x \in U : [x]_R \cap X \neq \phi\}$$

Now, if $R_*X = R^*X$, then X is called definable; otherwise, X is called a *rough set*.

Based on the equivalence relation on the universe of discourse, Dubois and Prade [6] introduced the lower and upper approximations of fuzzy sets in a Pawlak's approximation space [19] and obtained a new notion called rough fuzzy sets.

Definition 9.9 [6] Let (U, R) be a Pawlak's approximation space and $\mu \in I^U$. Then, the lower and upper rough approximations of μ in (U, R) are denoted by $\underline{R}(\mu)$ and $\bar{R}(\mu)$, respectively, which are fuzzy subsets in U defined by

$$\underline{R}(\mu)(x) = \wedge\{\mu(y) : y \in [x]_R\}$$

and

$$\bar{R}(\mu)(x) = \vee\{\mu(y) : y \in [x]_R\}, \quad \text{for all } x \in U.$$

The operators $\underline{R}$ and $\bar{R}$ are called the lower and upper rough approximation operators on fuzzy sets. μ is said to be definable in U if $\underline{R}(\mu) = \bar{R}(\mu)$; otherwise, μ is called a *rough fuzzy set*.

Feng et al. [9] provided a framework to combine rough sets and soft sets all together, which gives rise to several interesting new concepts such as soft rough sets and rough soft sets.

Definition 9.10 [9] Let $\Theta = (f, A)$ be a soft set over U. The pair $S = (U, \Theta)$ is called a soft approximation space. Based on S, the operators $\overline{apr}_S$ and $\underline{apr}_S$ are defined as:

$$\underline{apr}_S(X) = \{u \in U : \exists a \in A(u \in f(a) \subseteq X)\},$$

$$\underline{apr}_S(X) = \{u \in U : \exists a \in A(u \in f(a), f(a) \cap X \neq \phi)\}, \quad \text{for every } X \subseteq U$$

The two sets $\underline{apr}_S(X)$ and $\overline{apr}_S(X)$ are called the lower and upper soft rough approximations of X in S, respectively. If $\underline{apr}_S(X) = \overline{apr}_S(X)$, then X is said to be soft definable; otherwise, X is called a *soft rough set*.

Definition 9.11 [9] Let (U, R) be a Pawlak's approximation space and $\Theta = (f, A)$ be a soft set over U. Then, the lower and upper rough approximations of Θ in (U, R) are denoted by $R_*(\Theta) = (F_*, A)$ and $R^*(\Theta) = (F^*, A)$, respectively, which are soft sets over U defined by:

$$F_*(x) = \underline{R}(F(x)) = \{y \in U : [y]_R \subseteq F(x)\}$$

and

$$F^*(x) = \bar{R}(F(x)) = \{y \in U : [y]_R \cap F(x) \neq \phi\}$$

for all $x \in U$.

The operators R_* and R^* are called the lower and upper rough approximation operators on soft sets. If $R_*(\Theta) = R^*(\Theta)$, the soft set Θ is said to be definable; otherwise, Θ is called a *rough soft set*.

9.3 Soft Rough Fuzzy Sets and Soft Fuzzy Rough Soft Sets

Motivated by Dubois and Prade's original idea about rough fuzzy set, Feng et al. [9] introduced lower and upper soft rough approximations of fuzzy sets in a soft approximation space and obtained a new hybrid model called soft rough fuzzy set.

Definition 9.12 [9] Let $\Theta = (f, A)$ be a full soft set over U, i.e. $\cup_{a \in A} f(a) = U$ and the pair $S = (U, \Theta)$ be the soft approximation space. Then for a fuzzy set $\lambda \in I^U$, the lower and upper soft rough approximations of λ with respect to S are denoted by $\underline{sap}_S(\lambda)$ and $\overline{sap}_S(\lambda)$, respectively, which are fuzzy sets in U given by:

$$\underline{sap}_S(\lambda) = \{(x, \underline{sap}_S(\lambda)(x)) : x \in U\},$$
$$\overline{sap}_S(\lambda) = \{(x, \overline{sap}_S(\lambda)(x)) : x \in U\}, \text{ where}$$
$$\underline{sap}_S(\lambda)(x) = \wedge\{\mu_\lambda(y) : \exists a \in A \, (\{x, y\} \subseteq f(a))\} \text{ and}$$
$$\underline{sap}_S(\lambda)(x) = \vee\{\mu_\lambda(y) : \exists a \in A \, (\{x, y\} \subseteq f(a))\} \text{ for every } x \in U.$$

The operators $\underline{sap}_S$ and $\overline{sap}_S$ are called the lower and upper soft rough approximation operators on fuzzy sets. If $\underline{sap}_S(\lambda) = \overline{sap}_S(\lambda)$, then λ is said to be fuzzy soft definable; otherwise, λ is called a *soft rough fuzzy set*.

Meng et al. [17] introduced the lower and upper soft fuzzy rough approximations of a fuzzy set by granulating the universe of discourse with the help of a fuzzy soft set and obtained a new model called soft fuzzy rough set.

Definition 9.13 [17] Let $\Theta = (f, A)$ be a fuzzy soft set over U. Then, the pair $SF = (U, \Theta)$ is called a soft fuzzy approximation space. Then for a fuzzy set $\lambda \in I^U$, the lower and upper soft fuzzy rough approximations of λ with respect to SF are denoted by $\underline{\text{Apr}}_{SF}(\lambda)$ and $\overline{\text{Apr}}_{SF}$, respectively, which are fuzzy sets in U given by:

$$\underline{\text{Apr}}_{SF}(\lambda) = \{(x, \underline{\text{Apr}}_{SF}(\lambda)(x)) : x \in U\},$$

$$\overline{\text{Apr}}_{SF}(\lambda) = \left\{ \left(x, \underline{\text{Apr}}_{SF}(\lambda)(x) \right) : x \in U \right\}$$

where

$$\underline{\text{Apr}}_{SF}(\lambda)(x) = \wedge_{a \in A}((1 - f(a)(x)) \vee (\wedge_{y \in U}((1 - f(a)(y)) \vee \mu_\lambda(y))))$$

and $\overline{\text{Apr}}_{SF}(\lambda)(x) = \vee_{a \in A}(f(a)(x) \wedge (\vee_{y \in U}(f(a)(y) \wedge \mu_\lambda(y))))$ for every $x \in U$ and $\mu_\lambda(y)$ is the degree of membership of $y \in U$.

The operators $\underline{\text{Apr}}_{SF}$ and $\overline{\text{Apr}}_{SF}$ are called the lower and upper soft fuzzy rough approximation operators on fuzzy sets. If $\underline{\text{Apr}}_{SF}(\lambda) = \overline{\text{Apr}}_{SF}(\lambda)$, then λ is said to be soft fuzzy definable; otherwise, λ is called a *soft fuzzy rough set*.

9.4 Soft Interval-Valued Intuitionistic Fuzzy Rough Sets

In this section, we use an interval-valued intuitionistic fuzzy soft set to granulate the universe of discourse and obtain a new hybrid model called soft interval-valued intuitionistic fuzzy rough set.

Definition 9.14 Let us consider an interval-valued intuitionistic fuzzy set τ defined by $\tau = \{\langle x, \mu_\tau(x), \gamma_\tau(x) \rangle : x \in U\}$, where $\mu_\tau(x)$, $\gamma_\tau(x) \in \text{Int}([0, 1])$ for each $x \in U$ and $0 \leq \sup \mu_\tau(x) + \sup \gamma_\tau(x) \leq 1$.

Now let $\Theta = (f, A)$ be an interval-valued intuitionistic fuzzy soft set over U and $SIVIF = (U, \Theta)$ be the soft interval-valued intuitionistic fuzzy approximation space. Let $f: A \rightarrow \text{IVIFS}^U$ be defined by $f(a) = \{\langle x, \mu_{f(a)}(x), \gamma_{f(a)}(x) \rangle : x \in U\}$, for $a \in A$.

Then, the lower and upper soft interval-valued intuitionistic fuzzy rough approximations of τ with respect to SIVIF are denoted by $\downarrow\text{Apr}_{SIVIF}(\tau)$ and $\uparrow\text{Apr}_{SIVIF}(\tau)$, respectively, which are interval-valued intuitionistic fuzzy sets in U given by:

$$\downarrow \text{Apr}_{SIVIF}(\tau) = \left\{ \left\langle x, \left[\wedge_{a \in A}(\inf \mu_{f(a)}(x) \wedge \inf \mu_\tau(x)), \right. \right. \right.$$

$$\left. \wedge_{a \in A}(\sup \mu_{f(a)}(x) \wedge \sup \mu_\tau(x)) \right], \left[\wedge_{a \in A}(\inf \gamma_{f(a)}(x) \vee \inf \gamma_\tau(x)), \right.$$

$$\left. \left. \left. \wedge_{a \in A}(\sup \gamma_{f(a)}(x) \vee \sup \gamma_\tau(x)) \right] \right\rangle : x \in U \right\},$$

$$\uparrow\mathrm{Apr}_{\mathrm{SIVIF}}(\tau) = \Big\{ \Big\langle x, \Big[\wedge_{a\in A}(\inf \mu_{f(a)}(x) \vee \inf \mu_\tau(x)),$$

$$\wedge_{a\in A}(\sup \mu_{f(a)}(x) \vee \sup \mu_\tau(x))\Big], \Big[\wedge_{a\in A}(\inf \gamma_{f(a)}(x) \wedge \inf \gamma_\tau(x)),$$

$$\wedge_{a\in A}(\sup \gamma_{f(a)}(x) \wedge \sup \gamma_\tau(x))\Big]\Big\rangle : x \in U\Big\},$$

The operators $\downarrow\mathrm{Apr}_{\mathrm{SIVIF}}$ and $\uparrow\mathrm{Apr}_{\mathrm{SIVIF}}$ are called the lower and upper soft interval-valued intuitionistic fuzzy rough approximation operators on interval-valued intuitionistic fuzzy sets. If $\downarrow\mathrm{Apr}_{\mathrm{SIVIF}}(\tau) = \uparrow\mathrm{Apr}_{\mathrm{SIVIF}}(\tau)$, then τ is said to be soft interval-valued intuitionistic fuzzy definable; otherwise, τ is called a *soft interval-valued intuitionistic fuzzy rough set*.

It is to be noted that if $\mu_\tau(x)$, $\gamma_\tau(x) \in [0, 1]$ for each $x \in U$ and $0 \leq \mu_\tau(x) + \gamma_\tau(x) \leq 1$, then soft interval-valued intuitionistic fuzzy rough set becomes soft intuitionistic fuzzy rough set *and if* $\mu_\tau(x) \in [0, 1]$ with $\gamma_\tau(x) = 1 - \mu_\tau(x)$, then soft intuitionistic fuzzy rough set becomes soft fuzzy rough set.

Example 9.15 Let $U = \{x, y\}$ and $A = \{a, b\}$. Let (f, A) be an interval-valued intuitionistic fuzzy soft set over U, where $f: A \rightarrow \mathrm{IVIFS}^U$ be defined by

$$f(a) = \{\langle x, [0.2, 0.5], [0.3, 0.4]\rangle, \langle y, [0.6, 0.7], [0.1, 0.2]\rangle\},$$
$$f(b) = \{\langle x, [0.1, 0.3], [0.4, 0.5]\rangle, \langle y, [0.5, 0.8], [0.1, 0.2]\rangle\}.$$

Let $\tau = \{\langle x, [0.3, 0.4], [0.3, 0.4]\rangle, \langle y, [0.2, 0.4], [0.4, 0.5]\rangle\}$. Then

$$\downarrow\mathrm{Apr}_{\mathrm{SIVIF}}(\tau) = \{\langle x, [0.1, 0.3], [0.3, 0.4]\rangle, \langle y, [0.2, 0.4], [0.4, 0.5]\}\},$$
$$\uparrow\mathrm{Apr}_{\mathrm{SIVIF}}(\tau) = \{\langle x, [0.3, 0.4], [0.3, 0.4]\rangle, \langle y, [0.5, 0.7], [0.1, 0.2]\}\}.$$

Then, τ is a soft interval-valued intuitionistic fuzzy rough set.

Theorem 9.16 *Let* $\Theta = (f, A)$ *be an interval-valued intuitionistic fuzzy soft set over* U *and* $\mathrm{SIVIF} = (U, \Theta)$ *be the soft interval-valued intuitionistic fuzzy approximation space. Then,*

  (i) $\downarrow\mathrm{Apr}_{\mathrm{SIVIF}}(\phi) = \phi$
 (ii) $\uparrow\mathrm{Apr}_{\mathrm{SIVIF}}(U) = U$
(iii) $\downarrow\mathrm{Apr}_{\mathrm{SIVIF}}(\tau) \subseteq \tau \subseteq \uparrow\mathrm{Apr}_{\mathrm{SIVIF}}(\tau)$ *for* $\tau \in \mathrm{IVIFS}^U$.

Proof Straight forward.   □

Theorem 9.17 *Let* $\Theta = (f, A)$ *be an interval-valued intuitionistic fuzzy soft set over* U *and* $\mathrm{SIVIF} = (U, \Theta)$ *be the soft interval-valued intuitionistic fuzzy approximation space. Then, for* $\tau, \delta \in \mathrm{IVIFS}^U$, *we have*

  (i) $\tau \subseteq \delta \Rightarrow \uparrow\mathrm{Apr}_{\mathrm{SIVIF}}(\tau) \subseteq \downarrow\mathrm{Apr}_{\mathrm{SIVIF}}(\delta)$
 (ii) $\tau \subseteq \delta \Rightarrow \downarrow\mathrm{Apr}_{\mathrm{SIVIF}}(\tau) \subseteq \downarrow\mathrm{Apr}_{\mathrm{SIVIF}}(\delta)$
(iii) $\uparrow\mathrm{Apr}_{\mathrm{SIVIF}}(\tau \cap \delta) \subseteq \uparrow\mathrm{Apr}_{\mathrm{SIVIF}}(\tau) \cap \uparrow\mathrm{Apr}_{\mathrm{SIVIF}}(\delta)$
 (iv) $\downarrow\mathrm{Apr}_{\mathrm{SIVIF}}(\tau \cap \delta) \subseteq \downarrow\mathrm{Apr}_{\mathrm{SIVIF}}(\tau) \cap \downarrow\mathrm{Apr}_{\mathrm{SIVIF}}(\delta)$
  (v) $\uparrow\mathrm{Apr}_{\mathrm{SIVIF}}(\tau) \cup \uparrow\mathrm{Apr}_{\mathrm{SIVIF}}(\delta) \subseteq \uparrow\mathrm{Apr}_{\mathrm{SIVIF}}(\tau \cup \delta)$
 (vi) $\downarrow\mathrm{Apr}_{\mathrm{SIVIF}}(\tau) \cup \downarrow\mathrm{Apr}_{\mathrm{SIVIF}}(\delta) \subseteq \downarrow\mathrm{Apr}_{\mathrm{SIVIF}}(\tau \cup \delta)$

Proof (i)–(ii) are straight forward.

(iii) We have $\uparrow\mathrm{Apr}_{\mathrm{SIVIF}}(\tau \cap \delta)$

$$= \Big\{\Big\langle x, \Big[\wedge_{a\in A}\Big(\inf \mu_{f(a)}(x) \vee \inf \mu_{\tau\cap\delta}(x)\Big), \wedge_{a\in A}\Big(\sup \mu_{f(a)}(x) \vee \sup \mu_{t\cap d}(x)\Big)\Big],$$
$$\Big[\wedge_{a\in A}\Big(\inf \gamma_{f(a)}(x) \wedge \inf \gamma_{\tau\cap\delta}(x)\Big), \wedge_{a\in A}\Big(\sup \gamma_{f(a)}(x) \wedge \sup g_{\tau\cap\delta}(x)\Big)\Big]\Big\rangle : x \in U\Big\}$$

$$= \Big\{\Big\langle x, \Big[\wedge_{a\in A}\Big(\inf \mu_{f(a)}(x) \vee \min(\inf \mu_{\tau}(x), \inf \mu_{\delta}(x))\Big),$$
$$\wedge_{a\in A}\Big(\sup \mu_{f(a)}(x) \vee \min(\sup \mu_{\tau}(x), \sup \mu_{\delta}(x))\Big)\Big],$$
$$\Big[\wedge_{a\in A}\Big(\inf \gamma_{f(a)}(x) \wedge \max(\inf \gamma_{\tau}(x), \inf \gamma_{\delta}(x))\Big),$$
$$\wedge_{a\in A}\Big(\sup \gamma_{f(a)}(x) \wedge \max(\sup \gamma_{\tau}(x), \sup \gamma_{\delta}(x))\Big)\Big]\Big\rangle : x \in U\Big\}$$

Now $\uparrow\mathrm{Apr}_{\mathrm{SIVIF}}(\tau) \cap \uparrow\mathrm{Apr}_{\mathrm{SIVIF}}(\delta)$

$$= \Big\{\Big\langle x, [\min(\wedge_{a\in A}(\inf \mu_{f(a)}(x) \vee \inf \mu_{\tau}(x)),$$
$$\wedge_{a\in A}(\inf \mu_{f(a)}(x) \vee \inf \mu_{\delta}(x))),$$
$$\min(\wedge_{a\in A}(\sup \mu_{f(a)}(x) \vee \sup \mu_{\tau}(x)),$$
$$\wedge_{a\in A}(\sup \mu_{f(a)}(x) \vee \sup \mu_{\delta}(x)))],$$
$$[\max(\vee_{a\in A}(\inf \gamma_{f(a)}(x) \wedge \inf \gamma_{\tau}(x)),$$
$$\vee_{a\in A}(\inf \gamma_{f(a)}(x) \wedge \inf \gamma_{\delta}(x))),$$
$$\max(\wedge_{a\in A}(\sup \gamma_{f(a)}(x) \wedge \sup \gamma_{\tau}(x)),$$
$$\wedge_{a\in A}(\sup \gamma_{f(a)}(x) \wedge \sup \gamma_{\delta}(x))]\Big\rangle : x \in U\Big\}$$

As

$$\min(\inf \mu_{\tau}(x), \inf \mu_{\delta}(x)) \leq \inf \mu_{\tau}(x) \ \& \ \min(\inf \mu_{\tau}(x), \inf \mu_{\delta}(x)) \leq \inf \mu_{\delta}(x),$$

so

$$\wedge_{a\in A}(\inf \mu_{f(a)}(x) \vee \min(\inf \mu_{\tau}(x), \inf \mu_{\tau}(x))) \leq \wedge_{a\in A}(\inf \mu_{f(a)}(x) \vee \inf \mu_{\tau}(x))$$
$$\& \wedge_{a\in A}(\inf \mu_{f(a)}(x) \vee \min(\inf \mu_{\tau}(x), \inf \mu_{\tau}(x))) \leq \wedge_{a\in A}(\inf \mu_{f(a)}(x) \vee \inf \mu_{\tau}(x)).$$

Hence,

$$\wedge_{a\in A}\,(\inf \mu_{f(a)}(x) \vee \min(\inf \mu_\tau(x), \inf \mu_\delta(x))) \leq \min(\wedge_{a\in A}(\inf \mu_{f(a)}(x) \vee \inf \mu_\tau(x)),$$
$$\wedge_{a\in A}\,(\inf \mu_{f(a)}(x) \vee \inf \mu_\delta(x))).$$

Similarly,

$$\wedge_{a\in A}\,(\sup \mu_{f(a)}(x) \vee \min(\sup \mu_\tau(x),$$
$$\sup \mu_\delta(x))) \leq \min(\wedge_{a\in A}(\sup \mu_{f(a)}(x) \vee \sup \mu_\tau(x)),$$
$$\wedge_{a\in A}\,(\sup \mu_{f(a)}(x) \vee \sup \mu_\delta(x))).$$

Now, as

$$\max(\inf \gamma_\tau(x), \inf \gamma_\delta(x)) \geq \inf \gamma_\tau(x) \ \& \ \max(\inf \gamma_\tau(x), \inf \gamma_\delta(x)) \geq \inf \gamma_\delta(x),$$

we have

$$\wedge_{a\in A}\,(\inf \gamma_{f(a)}(x) \wedge \max(\inf \gamma_\tau(x), \inf \gamma_\delta(x))) \geq \wedge_{a\in A}\,(\inf \gamma_{f(a)}(x) \wedge \inf \gamma_\tau(x))$$
$$\& \ \wedge_{a\in A}\,(\inf \gamma_{f(a)}(x) \wedge \max(\inf \gamma_\tau(x), \inf \gamma_\delta(x))) \geq \wedge_{a\in A}\,(\inf \gamma_{f(a)}(x) \wedge \inf \gamma_\delta(x)).$$

Therefore,

$$\wedge_{a\in A}\,(\inf \gamma_{f(a)}(x) \wedge \max(\inf \gamma_\tau(x), \inf \gamma_\delta(x))) \geq \max(\wedge_{a\in A}(\inf \gamma_{f(a)}(x) \wedge \inf \gamma_\tau(x)),$$
$$\wedge_{a\in A}\,(\inf \gamma_{f(a)}(x) \wedge \inf \gamma_\delta(x))).$$

Similarly,

$$\wedge_{a\in A}\,(\sup \gamma_{f(a)}(x) \wedge \max(\sup \gamma_\tau(x),$$
$$\sup \gamma_\delta(x))) \geq \max(\wedge_{a\in A}(\sup \gamma_{f(a)}(x) \wedge \sup \gamma_\tau(x)),$$
$$\wedge_{a\in A}\,(\sup \gamma_{f(a)}(x) \wedge \sup \gamma_\delta(x))).$$

Consequently,

$$\uparrow\!\mathrm{Apr}_{\mathrm{SIVIF}}(\tau \cap \delta) \subseteq \uparrow\!\mathrm{Apr}_{\mathrm{SIVIF}}(\tau) \cap \uparrow\!\mathrm{Apr}_{\mathrm{SIVIF}}(\delta).$$

(iv) Proof is similar to (iii).
(v) We have $\uparrow\!\mathrm{Apr}_{\mathrm{SIVIF}}(\tau \cup \delta)$

$$= \left\{ \left\langle x, [\wedge_{a\in A}(\inf \mu_{f(a)}(x) \vee \inf \mu_{\tau\cup\delta}(x)), \right. \right.$$
$$\wedge_{a\in A}(\sup \mu_{f(a)}(x) \vee \sup \mu_{\tau\cup\delta}(x))],$$
$$[\wedge_{a\in A}(\inf \gamma_{f(a)}(x) \wedge \inf \gamma_{\tau\cup\delta}(x)),$$
$$\left. \left. \wedge_{a\in A}(\sup \gamma_{f(a)}(x) \wedge \sup \gamma_{\tau\cup\delta}(x))]\right\rangle x \in U\right\}$$
$$\left\{ \left\langle x, [\wedge_{a\in A}(\inf \mu_{f(a)}(x) \vee \max(\inf \mu_{\tau}(x), \inf \mu_{\delta}(x))), \right. \right.$$
$$\wedge_{a\in A}(\sup \mu_{f(a)}(x) \vee \max(\sup \mu_{\tau}(x), \sup \mu_{\delta}(x)))],$$
$$[\wedge_{a\in A}(\inf \gamma_{f(a)}(x) \wedge \min(\inf \gamma_{\tau}(x), \inf \gamma_{\delta}(x))),$$
$$\left. \left. \wedge_{a\in A}(\sup \gamma_{f(a)}(x) \wedge \min(\sup \gamma_{\tau}(x), \sup \gamma_{\delta}(x)))]\right\rangle : x \in U\right\}$$

Now $\uparrow\text{Apr}_{\text{SIVIF}}(\tau) \cup \uparrow\text{Apr}_{\text{SIVIF}}(\delta)$

$$= \left\{ \left\langle x, \left[\max(\wedge_{a\in A}(\inf \mu_{f(a)}(x) \vee \inf \mu_{\tau}(x)), \wedge_{a\in A}(\inf \mu_{f(a)}(x) \vee \inf \mu_{\delta}(x))), \right. \right. \right.$$
$$\left. \max(\wedge_{a\in A}(\sup \mu_{f(a)}(x) \vee \sup \mu_{\tau}(x)), \wedge_{a\in A}(\sup \mu_{f(a)}(x) \vee \sup \mu_{\delta}(x)))\right],$$
$$\left[\min(\wedge_{a\in A}(\inf \gamma_{f(a)}(x) \wedge \inf \gamma_{\tau}(x)), \wedge_{a\in A}(\inf \gamma_{f(a)}(x) \wedge \inf \gamma_{\delta}(x))), \right.$$
$$\min(\wedge_{a\in A}(\sup \gamma_{f(a)}(x) \wedge \sup \gamma_{\tau}(x)),$$
$$\left. \left. \left. \wedge_{a\in A}(\sup \gamma_{f(a)}(x) \wedge \sup \gamma_{\delta}(x)))\right]\right\rangle : x \in U\right\}$$

As

$$\max(\inf \mu_{\tau}(x), \inf \mu_{\delta}(x)) \geq \inf \mu_{\tau}(x) \ \& \ \max(\inf \mu_{\tau}(x), \inf \mu_{\delta}(x)) \geq \inf \mu_{\delta}(x),$$

so

$$\wedge_{a\in A}(\inf \mu_{f(a)}(x) \vee \max(\inf \mu_{\tau}(x), \inf \mu_{\delta}(x))) \geq \wedge_{a\in A}(\inf \mu_{f(a)}(x) \vee \inf \mu_{\tau}(x))$$
$$\& \ \wedge_{a\in A}(\inf \mu_{f(a)}(x) \vee \max(\inf \mu_{\tau}(x), \inf \mu_{\delta}(x))) \geq \wedge_{a\in A}(\inf \mu_{f(a)}(x) \vee \inf \mu_{\delta}(x)).$$

Hence,

$$\wedge_{a\in A}(\inf \mu_{f(a)}(x) \vee \max(\inf \mu_{\tau}(x), \inf \mu_{\delta}(x))) \geq \max(\wedge_{a\in A}(\inf \mu_{f(a)}(x) \vee \inf \mu_{\tau}(x)),$$
$$\wedge_{a\in A}(\inf \mu_{f(a)}(x) \vee \inf \mu_{\delta}(x))).$$

Similarly

$$\wedge_{a\in A}(\sup \mu_{f(a)}(x) \vee \max(\sup \mu_{\tau}(x), \sup \mu_{\delta}(x))) \leq \max(\wedge_{a\in A}(\sup \mu_{f(a)}(x) \vee \sup \mu_{\tau}(x)),$$
$$\wedge_{a\in A}(\sup \mu_{f(a)}(x) \vee \sup \mu_{\delta}(x))).$$

Now as

$$\min(\inf \gamma_\tau(x), \inf \gamma_\delta(x)) \leq \inf \gamma_\tau(x) \ \& \ \min(\inf \gamma_\tau(x), \inf \gamma_\delta(x)) \leq \inf \gamma_\delta(x),$$

we have

$$\wedge_{a \in A} (\inf \gamma_{f(a)}(x) \wedge \min(\inf \gamma_\tau(x), \inf \gamma_\delta(x))) \leq \wedge_{a \in A} (\inf \gamma_{f(a)}(x) \wedge \inf \gamma_\tau(x)) \ \&$$
$$\wedge_{a \in A} (\inf \gamma_{f(a)}(x) \wedge \min(\inf \gamma_\tau(x), \inf \gamma_\delta(x))) \leq \wedge_{a \in A} (\inf \gamma_{f(a)}(x) \wedge \inf \gamma_\delta(x)).$$

Therefore,

$$\wedge_{a \in A} (\inf \gamma_{f(a)}(x) \wedge \min(\inf \gamma_\tau(x), \inf \gamma_\delta(x))) \leq \min(\wedge_{a \in A}(\inf \gamma_{f(a)}(x) \wedge \inf \gamma_\tau(x)),$$
$$\wedge_{a \in A} (\inf \gamma_{f(a)}(x) \wedge \inf \gamma_\delta(x))).$$

Similarly,

$$\wedge_{a \in A} (\sup \gamma_{f(a)}(x) \wedge \min(\sup \gamma_\tau(x), \sup \gamma_\delta(x))) \leq \min(\wedge_{a \in A}(\sup \gamma_{f(a)}(x) \wedge \sup \gamma_\tau(x)),$$
$$\wedge_{a \in A} (\sup \gamma_{f(a)}(x) \wedge \sup \gamma_\delta(x))).$$

Consequently, $\uparrow Apr_{SIVIF}(\tau) \cup \uparrow Apr_{SIVIF}(\delta) \subseteq \uparrow Apr_{SIVIF}(\tau \cup \delta)$.
(vi) Proof is similar to (v).   □

9.5 A Multi-Criteria Group Decision-Making Problem

Example 9.18 Let $U = \{o_1, o_2, o_3, ..., o_r\}$ be a set of objects and E be a set of parameters and $A = \{e_1, e_2, e_3, ..., e_m\} \subseteq E$ and $\Theta = (F, A)$ be an interval-valued intuitionistic fuzzy soft set over U. Let us assume that we have an expert group $G = \{T_1, T_2, ..., T_n\}$ consisting of n specialists to evaluate the objects in U. Each specialist will examine all the objects in U and will point out his/her evaluation result. Let X_i denote the primary evaluation result of the specialist T_i. It is easy to see that the primary evaluation result of the whole expert group G can be represented as an interval-valued intuitionistic fuzzy evaluation soft set $S^* = (F^*, G)$ over U, where $F^*: G \to IVIFS^U$ is given by $F^*(T_i) = X_i$, for $i = 1, 2, ..., n$.

Now, we consider the soft interval-valued intuitionistic fuzzy rough approximations of the specialist T_i's primary evaluation result X_i with respect to the soft interval-valued intuitionistic fuzzy approximation space SIVIF $= (U, S)$. Then, we obtain two other interval-valued intuitionistic fuzzy soft sets $\downarrow S^* = (\downarrow F^*, G)$ and $\uparrow S^* = (\uparrow F^*, G)$ over U, where $\downarrow F^*: G \to IVIFS^U$ is given by $\downarrow F^*(T_i) = \downarrow Apr_{SIVIF}(X_i)$ and $\uparrow F^*: G \to IVIFS^U$ is given by $\uparrow F^*(T_i) = \uparrow Apr_{SIVIF}(X_i)$, for $i = 1, 2, ..., n$.

Here, $\downarrow S^*$ can be considered as the evaluation result for the whole expert group G with 'low confidence', $\uparrow S^*$ can be considered as the evaluation result for the whole expert group G with 'high confidence', and S^* can be considered as the evaluation result for the whole expert group G with 'middle confidence'.

Let us define two interval-valued intuitionistic fuzzy sets $\text{IVIFS}_{\downarrow S*}$ and $\text{IVIFS}_{\uparrow S*}$ by

$$\text{IVIFS}_{\downarrow S*} = \left\{ \left\langle o_k, \left[\frac{1}{n}\sum_{j=1}^{n}\inf \mu_{\downarrow F*(Tj)}(o_k), \frac{1}{n}\sum_{j=1}^{n}\sup \mu_{\downarrow F*(Tj)}(o_k), \right] \left[\frac{1}{n}\sum_{j=1}^{n}[\inf \gamma_{\downarrow F*(Tj)}(o_k), \right. \right. \right.$$
$$\left. \left. \left. \frac{1}{n}\sum_{j=1}^{n}\sup \gamma_{\downarrow F*(Tj)}(o_k)\right]\right\rangle : k = 1, 2, \ldots, r \right\}$$

and

$$\text{IVIFS}_{\uparrow S*} = \left\{ \left\langle o_k, \left[\frac{1}{n}\sum_{j=1}^{n}\inf \mu_{\uparrow F*(Tj)}(o_k), \frac{1}{n}\sum_{j=1}^{n}\sup \mu_{\uparrow F*(Tj)}(o_k), \right] \left[\frac{1}{n}\sum_{j=1}^{n}[\inf \gamma_{\uparrow F*(Tj)}(o_k), \right. \right. \right.$$
$$\left. \left. \left. \frac{1}{n}\sum_{j=1}^{n}\sup \gamma_{\uparrow F*(Tj)}(o_k)\right]\right\rangle : k = 1, 2, \ldots, r \right\}.$$

Now, we define another interval-valued intuitionistic fuzzy set IVIFS_S^* by

$$\text{IVIFS}_{S^*} = \left\{ \left\langle o_k, \left[\frac{1}{n}\sum_{j=1}^{n}\inf \mu_{F^*(T_j)}(o_k), \frac{1}{n}\sum_{j=1}^{n}\sup \mu_{F^*(T_j)}(o_k)\right], \right. \right.$$
$$\left. \left. \left[\frac{1}{n}\sum_{j=1}^{n}\inf \gamma_{F^*(T_j)}(o_k)\right], \left[\frac{1}{n}\sum_{j=1}^{n}\sup \gamma_{F^*(T_j)}(o_k)\right]\right\rangle : k = 1, 2, \ldots, r \right\}.$$

Then clearly, $\text{IVIFS}_{\downarrow S*} \subseteq \text{IVIFS}_{S*} \subseteq \text{IVIFS}_{\uparrow S*}$.

Let $C = \{L(\text{low confidence}), M(\text{middle confidence}), H(\text{high confidence})\}$ be a set of parameters. Let us consider the interval-valued intuitionistic fuzzy soft set $S^{**} = (f, C)$ over U, where $f: C \to \text{IVIFS}^U$ is given by

$$f(H) = \text{IVIFS}_{\uparrow S}^*, f(M) = \text{IVIFS}_S^*, f(L) = \text{IVIFS}_{\downarrow S}^*.$$

Now given a weighting vector $W = (w_L, w_M, w_H)$ such that $w_L, w_M, w_H \in \text{Int}([0, 1])$, we define $\alpha: U \to R^+$ by

$$\alpha(o_k) = \sup w_L * \sup \mu_{f(L)}(o_k) + \sup w_M * \sup \mu_{f(M)}(o_k) + \sup w_H * \sup \mu_{f(H)}(o_k), o_k \in U$$
(*represents ordinary multiplication).

Here, $\alpha(o_k)$ is called the weighted evaluation value of the alternative $o_k \in U$. Finally, we can select the object o_p such that $\alpha(o_p) = \max\{\alpha(o_k): k = 1, 2, \ldots, r\}$ as the most preferred alternative.

Algorithm

(1) Input the original description interval-valued intuitionistic fuzzy soft set (F, A).

(2) Construct the interval-valued intuitionistic fuzzy evaluation soft set $S^* = (F^*, G)$.

(3) Compute the soft interval-valued intuitionistic fuzzy rough approximations and then construct the interval-valued intuitionistic fuzzy soft sets $\downarrow S^*$ and $\uparrow S^*$.

(4) Construct the interval-valued intuitionistic fuzzy sets $IVIFS_{\downarrow S*}$, $IVIFS_{S*}$, $IVIFS_{\uparrow S*}$.

(5) Construct the interval-valued intuitionistic fuzzy soft set S^{**}.

(6) Input the weighting vector W and compute the weighted evaluation values $\alpha(o_k)$ of each alternative $o_k \in U$.

(7) Select the object o_p such that $\alpha(o_p) = \max\{\alpha(o_k): k = 1, 2,..., r\}$ as the most preferred alternative.

An illustrative example:

Let us consider a car selection problem to buy a car for the family of Mr. X. Let $U = \{c_1, c_2, c_3, c_4, c_5\}$ is the universe set consisting of five cars. Let us consider the soft set $S = (F, A)$, which describes the 'quality of the car', where $A = \{e_1(\text{expensive}), e_2(\text{fuel efficient}), e_3(\text{attractive}), e_4(\text{challenging internal structure with maximum seat capacity})\}$.

Let the tabular representation of the interval-valued intuitionistic fuzzy soft set (F, A) be Table 9.1.

Let $G = \{T_1, T_2, T_3, T_4, T_5\}$ be the set of members of the family of Mr. X to judge the quality of the car in U. Now, if X_i denote the primary evaluation result of the member T_i (for $i = 1, 2, 3, 4, 5$), then the primary evaluation result of the whole expert group G can be represented as an interval-valued intuitionistic fuzzy evaluation soft set $S^* = (F^*, G)$ over U, where $F^*: G \rightarrow IVIFS^U$ is given by $F^*(T_i) = X_i$ for $i = 1, 2, 3, 4, 5$.

Let the tabular representation of S^* be given as Table 9.2.

Table 9.1 Representation of the interval-valued intuitionistic fuzzy soft set (F, A)

	c_1	c_2	c_3	c_4	c_5
e_1	([0.2, 0.3], [0.4, 0.5])	([0.5, 0.7], [0.1, 0.3])	([0.4, 0.5], [0.2, 0.4])	([0.1, 0.2], [0.1, 0.3])	([0.3, 0.5], [0.3, 0.4])
e_2	([0.3, 0.6], [0.1, 0.2])	([0.1, 0.3], [0.2, 0.3])	([0.3, 0.6], [0.2, 0.4])	([0.5, 0.6], [0.2, 0.3])	([0.1, 0.3], [0.3, 0.6])
e_3	([0.4, 0.5], [0.2, 0.3])	([0.2, 0.4], [0.2, 0.5])	([0.1, 0.3], [0.4, 0.6])	([0.3, 0.4], [0.3, 0.4])	([0.4, 0.6], [0.1, 0.3])
e_4	([0.2, 0.4], [0.2, 0.4])	([0.6, 0.7], [0.1, 0.2])	([0.3, 0.4], [0.3, 0.4])	([0.2, 0.4], [0.4, 0.6])	([0.5, 0.7], [0.1, 0.2])

Table 9.2 Representation of the interval-valued intuitionistic fuzzy soft set S^*

	c_1	c_2	c_3	c_4	c_5
T_1	([0.4, 0.6], [0.1, 0.2])	([0.3, 0.4], [0.3, 0.4])	([0.2, 0.3], [0.2, 0.3])	([0.6, 0.8], [0.1, 0.2])	([0.1, 0.4], [0.2, 0.4])
T_2	([0.3, 0.5], [0.2, 0.4])	([0.5, 0.7], [0.1, 0.3])	([0.4, 0.6], [0.1, 0.3])	([0.3, 0.5], [0.1, 0.3])	([0.4, 0.5], [0.2, 0.3])
T_3	([0.1, 0.3], [0.5, 0.6])	([0.2, 0.3], [0.4, 0.5])	([0.1, 0.4], [0.2, 0.4])	([0.2, 0.3], [0.5, 0.6])	([0.3, 0.6], [0.2, 0.3])
T_4	([0.2, 0.3], [0.3, 0.4])	([0.4, 0.7], [0.1, 0.2])	([0.3, 0.5], [0.4, 0.5])	([0.4, 0.5], [0.2, 0.4])	([0.5, 0.7], [0.1, 0.2])
T_5	([0.6, 0.7], [0.1, 0.2])	([0.3, 0.5], [0.2, 0.5])	([0.5, 0.6], [0.3, 0.4])	([0.1, 0.3], [0.3, 0.6])	([0.1, 0.2], [0.6, 0.8])

Let us choose $P = (U, S)$ as the soft interval-valued intuitionistic fuzzy approximation space. Let us consider the interval-valued intuitionistic fuzzy evaluation soft sets $\downarrow S^* = (\downarrow F^*, G)$ and $\uparrow S^* = (\uparrow F^*, G)$ over U.

Then, after calculation we get the tabular representation of these sets as Tables 9.3 and 9.4.

Table 9.3 Representation of the interval-valued intuitionistic fuzzy soft set $\downarrow S^*$

	c_1	c_2	c_3	c_4	c_5
T_1	([0.2, 0.3], [0.1, 0.2])	([0.1, 0.3], [0.3, 0.4])	([0.1, 0.3], [0.2, 0.4])	([0.1, 0.2], [0.1, 0.3])	([0.1, 0.3], [0.2, 0.4])
T_2	([0.2, 0.3], [0.2, 0.4])	([0.1, 0.3], [0.1, 0.3])	([0.1, 0.3], [0.2, 0.4])	([0.1, 0.2], [0.1, 0.3])	([0.1, 0.3], [0.2, 0.3])
T_3	([0.1, 0.3], [0.5, 0.6])	([0.1, 0.3], [0.4, 0.5])	([0.1, 0.3], [0.2, 0.4])	([0.1, 0.2], [0.5, 0.6])	([0.1, 0.3], [0.2, 0.3])
T_4	([0.2, 0.3], [0.3, 0.4])	([0.1, 0.3], [0.1, 0.2])	([0.1, 0.3], [0.4, 0.5])	([0.1, 0.2], [0.2, 0.4])	([0.1, 0.3], [0.1, 0.2])
T_5	([0.2, 0.3], [0.1, 0.2])	([0.1, 0.3], [0.2, 0.5])	([0.1, 0.3], [0.3, 0.4])	([0.1, 0.2], [0.3, 0.6])	([0.1, 0.2], [0.6, 0.8])

Table 9.4 Representation of the interval-valued intuitionistic fuzzy soft set $\uparrow S^*$

	c_1	c_2	c_3	c_4	c_5
T_1	([0.4, 0.6], [0.1, 0.2])	([0.3, 0.4], [0.1, 0.2])	([0.2, 0.3], [0.2, 0.3])	([0.6, 0.8], [0.1, 0.2])	([0.1, 0.4], [0.1, 0.2])
T_2	([0.3, 0.5], [0.1, 0.2])	([0.5, 0.7], [0.1, 0.2])	([0.4, 0.6], [0.1, 0.3])	([0.3, 0.5], [0.1, 0.3])	([0.4, 0.5], [0.1, 0.2])
T_3	([0.2, 0.3], [0.1, 0.2])	([0.2, 0.3], [0.1, 0.2])	([0.1, 0.4], [0.2, 0.4])	([0.2, 0.3], [0.1, 0.3])	([0.3, 0.6], [0.1, 0.2])
T_4	([0.2, 0.3], [0.1, 0.2])	([0.4, 0.7], [0.1, 0.2])	([0.3, 0.5], [0.2, 0.4])	([0.4, 0.5], [0.1, 0.3])	([0.5, 0.7], [0.1, 0.2])
T_5	([0.6, 0.7], [0.1, 0.2])	([0.3, 0.5], [0.1, 0.2])	([0.5, 0.6], [0.2, 0.4])	([0.1, 0.3], [0.1, 0.3])	([0.1, 0.3], [0.1, 0.2])

Here, $\downarrow S^* \subseteq S^* \subseteq \uparrow S^*$.
Then we have,

$$\text{IVIFSet}_{\uparrow S^*} = \{\langle c_1, [0.34, 0.48], [0.10, 0.20]\rangle, \langle c_2, [0.34, 0.52], [0.10, 0.20]\rangle,$$
$$\langle c_3, [0.30, 0.48], [0.18, 0.36]\rangle, \langle c_4, [0.32, 0.48], [0.10, 0.28]\rangle,$$
$$\langle c_5, [0.28, 0.50], [0.10, 0.20]\rangle\},$$

$$\text{IVIFSet}_{\downarrow S^*} = \{\langle c_1, [0.18, 0.30], [0.24, 0.36]\rangle, \langle c_2, [0.10, 0.30], [0.22, 0.38]\rangle,$$
$$\langle c_3, [0.10, 0.30], [0.26, 0.42]\rangle, \langle c_4, [0.10, 0.20], [0.24, 0.44]\rangle,$$
$$\langle c_5, [0.10, 0.28], [0.26, 0.40]\rangle\},$$

$$\text{IVIFSet}_{S^*} = \{\langle c_1, [032, 0.48], [0.24, 0.36]\rangle, \langle c_2, [0.34, 0.52], [0.22, 0.38]\rangle,$$
$$\langle c_3, [0.30, 0.48], [0.24, 0.38]\rangle, \langle c_4, [0.32, 0.48], [0.24, 0.42]\rangle,$$
$$\langle c_5, [0.28, 0.48], [0.26, 0.40]\rangle\},$$

Thus,

$$\text{IVIFS}^*_{\downarrow S} \subseteq \text{IVIFSet}^*_S \subseteq \text{IVIFS}^*_{\uparrow S}$$

Let $C = \{L(\text{low confidence}), M(\text{middle confidence}), H(\text{high confidence})\}$ be a set of parameters. Let us consider the interval-valued intuitionistic fuzzy soft set $S^{**} = (f, C)$ over U, where $f: C \rightarrow \text{IVIFS}^U$ is given by

$$f(H) = \text{IVIFS}^*_S, f(M) = \text{IVIFS}^*_S, f(L) = \text{IVIFS}^*_{\downarrow S}.$$

Now assuming the weighting vector $W = (w_L, w_M, w_H)$ such that $w_L = [0.5, 0.6]$, $w_M = [0.4, 0.5]$, $w_H = [0.4, 0.7]$, we have,

$$\alpha(c_1) = 0.6 \times 0.30 + 0.5 \times 0.48 + 0.7 \times 0.48 = 0.756,$$
$$\alpha(c_2) = 0.6 \times 0.30 + 0.5 \times 0.52 + 0.7 \times 0.52 = 0.804,$$
$$\alpha(c_3) = 0.6 \times 0.30 + 0.5 \times 0.48 + 0.7 \times 0.48 = 0.756,$$
$$\alpha(c_4) = 0.6 \times 0.20 + 0.5 \times 0.48 + 0.7 \times 0.48 = 0.696,$$
$$\alpha(c_5) = 0.6 \times 0.28 + 0.5 \times 0.48 + 0.7 \times 0.50 = 0.758.$$

As $\max\{\alpha(c_1), \alpha(c_2), \alpha(c_3), \alpha(c_4), \alpha(c_5)\} = 0.804$, so the car c_2 will be selected as the most preferred alternative.

9.6 Conclusion

In this chapter, we first defined soft interval-valued intuitionistic fuzzy rough set which are the extension of soft intuitionistic fuzzy rough set and soft fuzzy rough set. We also investigated some basic properties of soft interval-valued intuitionistic fuzzy rough set. Finally, we have proposed a soft interval-valued intuitionistic fuzzy rough set-based multi-criteria group decision-making scheme and presented an example regarding the car selection problem for a family to buy a car to show that this scheme successfully works. It is to be noted that we defined soft interval-valued intuitionistic fuzzy rough set in such a way so that complicated calculations in decision-making problems will be avoided.

References

1. Aktas, H., Cagman, N.: Soft sets and soft groups. Inf. Sci. **177**(13), 2726–2735 (2007)
2. Ali, M.I., Feng, F., Liu, X., Min, W.K., Shabir, M.: On some new operations in soft set theory. Comput. Math. Appl. **57**, 1547–1553 (2009)
3. Atanassov, K.: Intuitionistic fuzzy sets. Fuzzy Sets Syst. **20**, 87–96 (1986)
4. Atanassov, K., Gargov, G.: Interval valued intuitionistic fuzzy sets. Fuzzy Sets Syst. **31**, 343–349 (1989)
5. Chen, D., Tsang, E.C.C., Yeung, D.S., Wang, X.: The parameterization reduction of soft Sets and its applications. Comput. Math. Appl. **49**, 757–763 (2005)
6. Dubois, D., Prade, H.: Rough fuzzy sets and fuzzy rough sets. Int. J. Gen Syst. **17**, 191–209 (1990)
7. Feng, F., Jun, Y.B., Zhao, X.: Soft semi rings. Comput. Math. Appl. **56**, 2621–2628 (2008)
8. Feng, F., Jun, Y.B., Liu, X., Li, L.: An adjustable approach to fuzzy soft set based decision making. J. Comput. Appl. Math. **234**, 10–20 (2010)
9. Feng, F., Li, C.X., Davvaz, B., Ali, M.I.: Soft sets combined with fuzzy sets and rough sets: a tentative approach. Soft. Comput. **14**, 899–911 (2010)
10. Feng, F.: Soft rough sets applied to multi criteria group decision making. Ann. Fuzzy Math. Inf. **2**, 69–80 (2011)
11. Jiang, Y., Tang, Y, Chen, Q.,Liu, H.,Tang, J.: Interval valued intuitionistic fuzzy soft sets and their properties. Comput. Math. Appl. **60**, 906–918 (2010)
12. Jun, Y.B.: Soft BCK/BCI-algebras. Comput. Math. Appl. **56**, 1408–1413 (2008)
13. Kong, Z., Gao, L., Wang, L., Li, S.: The normal parameter reduction of soft sets and its algorithm. Comput. Math. Appl. **56**, 3029–3037 (2008)
14. Maji, P.K., Biswas, R., Roy, A.R.: Soft set theory. Comput. Math. Appl. **45**, 555–562 (2003)
15. Maji, P.K., Biswas, R., Roy, A.R.: Fuzzy soft sets. J. Fuzzy Math. **9**, 589–602 (2001)
16. Maji, P.K., Biswas, R., Roy, A.R.: Intuitionistic fuzzy soft sets. J. Fuzzy Math. **12**, 677–692 (2004)
17. Meng, D., Zhang, X., Qin, K.: Soft rough fuzzy sets and soft fuzzy rough sets. Comput. Math. Appl. **62**, 4635–4645 (2011)
18. Molodtsov, D.: Soft set theory-first results. Comput. Math. Appl. **37**, 19–31 (1999)
19. Pawlak, Z.: Rough sets. Int. J. Comput. Inf. Sci. **11**, 341–356 (1982)
20. Roy, A.R., Maji, P.K.: A fuzzy soft set theoretic approach to decision making problems. J. Comput. Appl. Math. **203**, 412–418 (2007)
21. Sun, Q.M., Zhang, Z. L., Liu, J.: Soft sets and soft modules. In: Wang, G., Li, T., Grzymala-Busse, J.W., Miao, D., Skowron, A., Yao Y., (eds.) Proceedings of the Third International Conference on Rough Sets and Knowledge Technology, RSKT 2008. Lecture notes in Computer Science, vol. 5009, pp. 403–409. Springer, Berlin (2008)
22. Xiao, Z., Gong, K., Zou, Y.: A combined forecasting approach based on fuzzy soft sets. J. Comput. Appl. Math. **228**, 326–333 (2009)
23. Yang, X.B., Lin, T.Y., Yang, J.Y., Li, Y., Yu, D.J.: Combination of interval valued fuzzy set and soft set. Comput. Math. Appl. **58**, 521–527 (2009)

24. Yang, X.B., Yu, D.J., Yang, J.Y., Wu, C.: Generalization of soft set theory: from crisp to fuzzy case. In: Cao B.Y. (ed.) Proceeding of the Second International Conference on Fuzzy Information and Engineering. Advances on Soft Computing, vol. 40, pp. 345–354. Springer, Berlin (2007)
25. Yao, Y.Y.: A comparative study of fuzzy sets and rough sets. Inf. Sci. **109**, 227–242 (1998)
26. Zadeh, L.A.: Fuzzy sets. Inf. Control **8**, 338–353 (1965)

Chapter 10
IF Parameterised Intuitionistic Fuzzy Soft Set Theories on Decisions-Making

Abstract In this chapter, we introduce the concept of intuitionistic fuzzy parameterised intuitionistic fuzzy soft (ifpifs) sets and their operations with examples. We also define the approximate functions of ifpifs-set from the intuitionistic fuzzy parameterised set to the intuitionistic fuzzy subsets [1] of universal set. Lastly, we construct an ifpifs-set decision-making problem and try to solve the problem.

Keywords Fuzzy soft set · Intuitionistic fuzzy set · Intuitionistic fuzzy soft set

In our real-life problems, there are situations with the uncertain data that may not be successfully modelled by the classical mathematics. There are some mathematical tools for dealing with uncertainties—they are fuzzy set theory introduced by Zadeh [12] and soft set theory initiated by Molodtsov [10] further studied by Aktas and Cagman [2]. The concepts have been generalised in [4–9, 11] in fuzzy setting, which are related to our work. The aim of this chapter is to construct ifpifs-set decision-making problem and to solve the problem with IF parameterised intuitionistic fuzzy soft set theories.

Definition 10.1 Let U be an initial universe, E be the set of IF parameters and X be a IF set over E with the membership function $\mu_X: E \to [0,1]$ and non-membership function $\gamma_X: E \to [0,1]$ where $0 \le \mu_X(x) + \gamma_X(x) \le 1$ and $\eta_X(x) = \{(A_X(x), B_X(x))/u\}$ be an IF set over U for all $x \in E$; $A_X(x), B_X(x) \in [0,1]$; and $u \in U$. Then, an ifpifs-set Γ_X over U is a set defined by a function $\eta_X(x)$ represents a mapping $\eta_X: E \to P(U)$ such that $\eta_X(x) = \phi$ if $\mu_X(x) = 0$ and $\gamma_X(x) = 1$. Here, η_X is called the IF approximate function of the ifpifs-set Γ_X and the value $\eta_X(x)$ is an IF set called x-element of the ifpifs-set for all $x \in E$. Thus, an ifpifs set Γ_X over U can be represented by the set of ordered pairs

$$\Gamma_X = \{((\mu_X(x), \gamma_X(x))/x), \eta_X(x)) : x \in E, \eta_X(x) \in P(U), \mu_X(x), \gamma_X(x) \in [0,1]\}.$$
$$= \{(((\mu_X(x), \gamma_X(x))/x), ((A_X(x), B_X(x))/u)\} : x \in E, \eta_X(x)$$
$$= (A_X(x), B_X(x))/u) \in P(U), \mu_X(x), \gamma_X(x), A_X(x), B_X(x) \in [0,1]\}.$$

$P(U)$ is the family of intuitionistic fuzzy subsets of U. We denote the sets of all ifpifs-sets over U by IFPIFS(U).

© Springer India 2015
A. Mukherjee, *Generalized Rough Sets*, Studies in Fuzziness
and Soft Computing 324, DOI 10.1007/978-81-322-2458-7_10

Example 10.2 Let $U = \{u_1, u_2, u_3, u_4, u_5\}$ is a universal set and $E = \{x_1, x_2, x_3, x_4\}$ is a set of IF parameters. If $X = \{(0.2, 0.7)/x_2, (0.5, 0.3)/x_3, (1, 0)/x_4\}$ and $\eta_X(x_2) = \{(0.5, 0.4)/u_1, (0.3, 0.6)/u_3\}$, $\eta_X(x_3) = \phi$, $\eta_X(x_4) = U$, then the ipifs-set Γ_X is written as

$$\Gamma_X = \{(0.2, 0.7)/x_1, \{(0.5, 0.4)/u_1, (0.3, 0.6)/u_3\}), ((1, 0)/x_4, U)\}.$$

Definition 10.3 Let $\Gamma_X \in \text{IFPIFS}(U)$. If, $\eta_X(x) = U$ for all $x \in E$, i.e., $\mu_X(x) = 1$, $\gamma_X(x) = 0$, then Γ_X is called X-universal ipifs-set, denoted by Γ_X.

If $X = E$, then the X-universal ipifs-set (Γ_X) is called universal ipifs-set denoted by Γ_E.

Definition 10.4 Let $\Gamma_X \in \text{IFPIFS}(U)$. If $\eta_X(x) = \Phi$ for all $x \in E$, i.e., $\mu_X(x) = 0$, $\gamma_X(x) = 1$, then Γ_X is called X-empty ipifs-set, denoted by $\Gamma_{\Phi X}$.

If $X = \Phi$ then the X-empty ipifs-set ($\Gamma_{\Phi X}$) is called empty ipifs-set denoted by Γ_Φ.

Example 10.5 Let $U = \{u_1, u_2, u_3, u_4, u_5\}$ is a universal set and $E = \{x_1, x_2, x_3, x_4\}$ is a set of IF parameters. If $X = \{(0.2, 0.7)/x_2, (0.5, 0.3)/x_3, (1, 0)/x_4\}$ and $\eta_X(x_2) = \{(0.5, 0.4)/u_1, (0.3, 0.6)/u_3\}$, $\eta_X(x_3) = \phi$, $\eta_X(x_4) = U$, then the ipifs-set Γ_X is written as $\Gamma_X = \{(0.2, 0.7)/x_2, \{(0.5, 0.4)/u_1, (0.3, 0.6)/u_3\}), ((1, 0)/x_4, U)\}$. Now, if $Y = \{(1, 0)/x_1, (0.7, 0.2)/x_4\}$ and $\eta_X(x_1) = \Phi$, $\eta_X(x_4) = \Phi$, then ipifs-set Γ_Y is a Y-empty ipifs-set, i.e., $\Gamma_Y = \Gamma_{\Phi Y}$.

If $Z = \{(1, 0)/x_1, (1, 0)/x_2), \eta_X(x_1) = U, \eta_X(x_2) = U$, then the ipifs-set Γ_Z is Z-universal ipifs-set.

If $X = \Phi$, then ipifs-set Γ_X is an empty set i.e., $\Gamma_X = \Gamma_\Phi$.

If $X = E$ and $\eta_X(x_i) = U$ for all $x_i \in E$ ($i = 1, 2, 3, 4$), then ipifs-set Γ_X is a universal ipifs-set, i.e., $\Gamma_X = \Gamma_E$.

Definition 10.6 Let $\Gamma_X, \Gamma_Y \in \text{IFPIFS}(U)$. Then, Γ_X is an ipifs-subset of Γ_Y, denoted by $\Gamma_X \subseteq \Gamma_Y$ if $\mu_X(x) \leq \mu_Y(x)$, $(\gamma_X(x) \geq \gamma_Y(x)$, and $\eta_X(x) = ((A_X(x), B_X(x))/u) \subseteq \eta_Y(x) = ((A_Y(x), B_Y(x))/u)$ for all $x \in E$.

Definition 10.7 Let $\Gamma_X, \Gamma_Y \in \text{IFPIFS}(U)$. Then, Γ_X and Γ_Y are ipifs-equal of Γ_Y, written as $\Gamma_X = \Gamma_Y$ if $\mu_X(x) = \mu_Y(x)$, $(\gamma_X(x) = \gamma_Y(x)$, and $\eta_X(x) = ((A_X(x), B_X(x))/u) = \eta_Y(x) = ((A_Y(x), B_Y(x))/u)$ for all $x \in E$.

Definition 10.8 Let $\Gamma_X, \Gamma_Y, \Gamma_Z \in \text{IFPIFS}(U)$. Then,

(i) $(\Gamma_X = \Gamma_Y$ and $\Gamma_Y = \Gamma_Z) \Rightarrow \Gamma_X = \Gamma_Z$.

(ii) $(\Gamma_X \subseteq \Gamma_Y$ and $\Gamma_Y \subseteq \Gamma_X) \Rightarrow \Gamma_X = \Gamma_Y$.

Proof The proofs are trivial.   □

Definition 10.9 Let $\Gamma_X \in \text{IFPIFS}(U)$, then the complement of Γ_X denoted by Γ_X^c is defined by

$$\Gamma_X = \{(\gamma_X(x), \mu_X(x)/x, \eta_X^c(x)) : x \in E, \eta_X^c(x) \in P(U), \mu_X(x), \gamma_X(x) \in [0, 1]\},$$

where

$$\eta_X^c = (\eta_X(x))^c = U - \eta_X(x) = \{U - (A_X(x), B_X(x))/u\}$$
$$= U \cap (B_X(x), A_X(x))/u$$
$$= (B_X(x), A_X(x))/u.$$

Definition 10.10

(a) Let $\Gamma_X, \Gamma_Y \in \text{IFPIFS}(U)$. Then, union of Γ_X and Γ_Y, denoted by $\Gamma_X \cup \Gamma_Y$, is defined by

$$\Gamma_X \cup \Gamma_Y = \{(\mu_{X\cup Y}(x), \gamma_{X\cap Y}(x)/x), \eta_{X\cup Y}(x)\}$$
$$= [(\max\{\mu_X(x), \mu_Y(x)\}, \min\{\gamma_X(x), \gamma_Y(x)\})/x,$$
$$(\eta_X(x) \cup \eta_Y(x))]$$
$$= [(\max\{\mu_X(x), \mu_Y(x)\}, \min\{\gamma_X(x), \gamma_Y(x)\})/x,$$
$$\{(A_X(x), B_X(x))/u \cup (A_Y(x), B_Y(x))/u\}].$$
$$= [(\max\{\mu_X(x), \mu_Y(x)\}, \min\{\gamma_X(x), \gamma_Y(x)\})/x, \{\max(A_X(x), A_Y(x)),$$
$$\min(B_X(x), B_Y(x))\}/u], \quad \text{for all } x \in E.$$

(b) The intersection of Γ_X and Γ_Y, denoted by $\Gamma_X \cap \Gamma_Y$, is defined by

$$\Gamma_X \cap \Gamma_Y = \{(\mu_{X\cap Y}(x), \gamma_{X\cup Y}(x)/x), \eta_{X\cap Y}(x)\}$$
$$= [(\min\{\mu_X(x), \mu_Y(x)\}, \max\{\gamma_X(x), \gamma_Y(x)\})/x, (\eta_X(x) \cap \eta_Y(x))]$$
$$= [(\min\{\mu_X(x), \mu_Y(x)\}, \max\{\gamma_X(x), \gamma_Y(x)\})/x,$$
$$\{(A_X(x), B_X(x))/u \cap (A_Y(x), B_Y(x))/u\}].$$
$$= [(\min\{\mu_X(x), \mu_Y(x)\}, \max\{\gamma_X(x), \gamma_Y(x)\})/x,$$
$$\{\min(A_X(x), A_Y(x)), \max(B_X(x), B_Y(x))\}/u,] \quad \text{for all } x \in E.$$

Result 10.11 Let $\Gamma_X, \Gamma_Y, \Gamma_Z \in \text{IFPIFS}(U)$. Then,

(i) $\Gamma_X \cup \Gamma_X = \Gamma_X$ and $\Gamma_X \cap \Gamma_X = \Gamma_X$.
(ii) $\Gamma_{\Phi X} \cup \Gamma_X = \Gamma_X$ and $\Gamma_{\Phi X} \cap \Gamma_X = \Gamma_X$.
(iii) $\Gamma_\Phi \cup \Gamma_X = \Gamma_X$ and $\Gamma_\Phi \cap \Gamma_X = \Gamma_\Phi$.
(iv) $\Gamma_X \cup \Gamma_E = \Gamma_E$ and $\Gamma_X \cap \Gamma_E = \Gamma_X$.
(v) $\Gamma_X \cup \Gamma_Y = \Gamma_Y \cup \Gamma_X$ and $\Gamma_X \cap \Gamma_Y = \Gamma_Y \cap \Gamma_X$.
(vi) $(\Gamma_X \cup \Gamma_Y) \cup \Gamma_Z = \Gamma_X \cup (\Gamma_Y \cup \Gamma_Z)$ and $(\Gamma_X \cap \Gamma_Y) \cap \Gamma_Z = \Gamma_X \cap (\Gamma_Y \cap \Gamma_Z)$.

It is to be noted that if $\Gamma_X \neq \Gamma_E$ or $\Gamma_X \neq \Gamma_\Phi$, then $\Gamma_X \cup \Gamma_X^c \neq \Gamma_E$ and $\Gamma_X \cap \Gamma_X^c \neq \Gamma_\Phi$.

Theorem 10.12 *Let $\Gamma_X, \Gamma_Y \in$ IFPIFS(U). Then, De Morgan's laws are valid*

(i) $(\Gamma_X \cup \Gamma_Y)^c = \Gamma_X^c \cap \Gamma_Y^c.$
(ii) $(\Gamma_X \cap \Gamma_Y)^c = \Gamma_X^c \cup \Gamma_Y^c.$

Proof (i) For all $x \in E$,

$$\left(\mu_{(X \cup Y)}^c(x), \gamma_{(X \cap Y)}^c(x)\right) = \{1 - \mu_{X \cup Y}(x), 1 - \gamma_{X \cap Y}(x)\}$$
$$= \{(1 - \max[\mu_X(x), \mu_Y(x)])(1 - \min[\gamma_X(x, \gamma_Y(x)])\}$$
$$= \min\{1 - \mu_X(x), 1 - \mu_Y(x)\}, \max\{1 - \gamma_X(x), 1 - \gamma_Y(x)\}$$
$$= \min\{\mu_X^c(x), \mu_Y^c(x)\}, \max\{\gamma_X^c(x), \gamma_Y^c(x)\}$$
$$= \mu_X^c \cap_Y^c (x), \gamma_X^c \cup_Y^c (x)$$

and

$$\eta_{(X \cup Y)}^c(x) = \eta_{(X \cup Y)}^c(x)$$
$$= (\eta_X(x) \cup \eta_Y(x))^c$$
$$= (\eta_X(x))^c \cap (\eta_Y(x))^c$$
$$= \eta_X^c(x) \cap \eta_Y^c(x)$$
$$= (B_X(x), A_X(x)/u) \cap (B_Y(x), A_Y(x)/u)$$
$$= \{\min(B_X(x), B_Y(x)), \max(A_X(x), A_Y(x))\}/u$$
$$= (B_{X \cap Y}(x), A_{X \cup Y}(x))/u$$
$$= \eta_X^c \cap_Y^c (x).$$

Hence the result.
    Likewise, the proof of (ii) can be made easily.   □

Theorem 10.13 *Let $\Gamma_X, \Gamma_Y, \Gamma_Z \in$ IFPIFS(U). Then,*

(i) $\Gamma_X \cup (\Gamma_Y \cap \Gamma_Z) = (\Gamma_X \cup \Gamma_Y) \cap (\Gamma_X \cup \Gamma_Z).$
(ii) $\Gamma_X \cap (\Gamma_Y \cup \Gamma_Z) = (\Gamma_X \cap \Gamma_Y) \cup (\Gamma_X \cap \Gamma_Z).$

Proof For all $x \in E$,

$$\left(\mu_{(X \cup (Y \cap Z))}(x), \gamma_{(X \cap (Y \cup Z))}(x)\right) = \{(\max[\mu_X(x), \mu_{Y \cap Z}(x)]), ([\min \gamma_X(x, \gamma_{Y \cup Z}(x)])\}$$
$$= \max\{\mu_X(x), \min(\mu_Y(x), \mu_Z(x))\},$$
$$\min\{\gamma_X(x), \max(\gamma_Y(x), \eta_Z(x))\}$$
$$= \min\{\max(\mu_X(x), \mu_Y(x)), \max(\mu_X(x), \mu_Z(x))\},$$
$$\max\{\min(\gamma_X(x), \gamma_Y(x)), \min(\gamma_X(x), \gamma_Z(x))\}$$
$$= \mu_{(X \cup Y) \cap (X \cup Z)}(x), \gamma_{(X \cap Y) \cup (X \cap Z)}(x)$$

and

$$\begin{aligned}
\eta_{(X \cup (Y \cap Z))}(x) &= \eta_X(x) \cup \eta_{(Y \cap Z)}(x) \\
&= \eta_X(x) \cup \{\eta_Y(x) \cap \eta_Z(x)\} \\
&= (\eta_X(x) \cup \eta_Y(x)) \cap (\eta_X(x) \cup \eta_Z(x)) \\
&= (\eta_{X \cup Y}(x) \cap \eta_{X \cup Z}(x)) \\
&= (\eta_{(X \cup Y) \cap (X \cup Z)})(x).
\end{aligned}$$

Likewise, the proof of (ii) can be made in a similar way.

10.1 ifpifs-Aggregation Operator

In this section, we define an aggregate IF set of an ifpifs-set. We also define ifpifs-aggregation operator that produce an aggregate IF set from an ifpifs-set and its IF parameter set.

The concept of IF parameterised intuitionistic fuzzy soft set (ifpifs-set) and their operations are given in Sect. 10.1.

We define an aggregate IF set of an ifpifs-set. We also define ifpifs-aggregation operator that produce an aggregate IF set from an ifpifs-set and its IF parameter set.

Definition 10.14 Let $\Gamma_X \in$ IFPIFS(U); then, ifpifs-aggregation operator, denoted by IFPIFSagg, is defined by IFPIFSagg: $P(E) \times$ IFPIFS(U) $\to P(U)$ where IFPIFSagg $(X, \Gamma_X) = \Gamma_X^*$ and $\Gamma_X^* = \{\mu \Gamma_X^*(u), \gamma \Gamma_X^*(u)/u: u \in U\}$, which is an IF set over U. The value Γ_X^* is called aggregate IF set of the set Γ_X. Here, the membership degree $\mu \Gamma_X^*(u)$ of u and the non-membership degree of $\gamma \Gamma_X^*(u)$ of u are defined as

$$\left(\mu \Gamma_X^*(u), \gamma \Gamma_X^*(u)\right) = 1/|E| \sum_{x \in E} \{\mu_X(x) A_X(x), \gamma_X(x) B_X(x)\}$$

where $\eta_X(x) = (A_X(x), B_X(x))/u$ and $|E|$ is the cardinality of E.

We now construct an ifpifs-set decision-making method by the following steps:

(i)   First construct an ifpifs-set Γ_X over U.
(ii) Find the aggregate IF set Γ_X^* of Γ_X.
(iii) Find the maximum membership grade of $\mu \Gamma_X^*(u)$, and observe the values of $\gamma \Gamma_X^*(u)$.

Example 10.15 Now we give an example for the above concept. Assume that an office wants to file a post, there are eight candidates. So, $U = \{u_1, u_2, u_3, u_4, u_5, u_6, u_7, u_8\}$. The recruiting committee considers a set of parameters $E = \{x_1, x_2, x_3, x_4, x_5\}$. The parameters x_i, ($i = 1, 2, 3, 4, 5$) stand for 'experience', 'computer knowledge', 'young age', 'good speaking', and 'friendly', respectively. After the interview, each candidate is evaluated from point of view of goals and the constraint according to a chosen subset $X = \{(0.5, 0.4)/x_2, (0.9, 0.1)/x_3, (0.6, 0.3)/x_4\}$ of E, and finally, the committee constructs the following ifpifs-set over U $\Gamma_X = [((0.5, 0.4)/x_2, \{(0.3, 0.6)/u_2, (0.4, 0.5)/u_3, (0.1, 0.9)/u_4, (0.9, 0.1)/u_5, (0.7, 0.2)/u_7\}), ((0.9, 0.1)/x_3, \{(0.4, 0.5)/u_1, (0.4, 0.6)/u_2, (0.9, 0.1)/u_3, (0.3, 0.6)/u_4\}), ((0.6, 0.3)/x_4, \{(0.2, 0.7)/u_1, (0.5, 0.4)/u_2, (0.1, 0.8)/u_5, (0.7, 0.3)/u_7, (1, 0)/u_8\})]$.

Thus, the aggregate IF set $\Gamma_X^* = \{(0.096, 0.052)/u_1, (0.162, 0.084)/u_2, (0.202, 0.042)/u_3, (0.064, 0.084)/u_4, (0.102, 0.056)/u_5, (0.154, 0.034)/u_7, (0.12, 0)/u_8\}$.

Finally the largest membership-grade $\mu\Gamma_X^*(u) = 0.202$ has been chosen, and it is to be observed that the corresponding value of $\gamma\Gamma_X^*(u) = 0.042$, which means that the candidate u_3 is selected for the post. Now the question arises 'what is the role of $\gamma\Gamma_X^*(u)$'? Here is the answer—In case the largest membership-grade $\mu\Gamma_X^*(u)$ are same for some candidates, then chose the candidate having the smallest non-membership-grade $\gamma\Gamma_X^*(u)$ among them.

Example 10.16 Considering the Example 10.2, we construct the ifpifs-set over U as $\Gamma_X = [((0.5, 0.4)/x_2, \{(0.4, 0.6)/u_2, (0.4, 0.5)/u_3, (0.1, 0.9)/u_4, (0.9, 0.1)/u_5, (0.7, 0.2)/u_7\}), ((0.9, 0.1)/x_3, \{(0.4, 0.5)/u_1, (0.4, 0.6)/u_2, (0.9, 0.1)/u_3, (0.3, 0.6)/u_4\}), ((0.6, 0.3)/x_4, \{(0.2, 0.7)/u_1, (0.6, 0.4)/u_2, (0.1, 0.8)/u_5, (0.7, 0.3)/u_7, (1, 0)/u_8\})]$.

Then, the aggregate IF set $\Gamma_X^* = \{(0.096, 0.052)/u_1, (0.202, 0.084)/u_2, (0.202, 0.042)/u_3, (0.064, 0.084)/u_4, (0.102, 0.056)/u_5, (0.154, 0.034)/u_7, (0.12, 0)/u_8\}$.

Here, u_2 and u_3 have the same largest membership-grade $\mu\Gamma_X^*(u) = 0.202$, but we observe that the minimum non-membership value $\gamma\Gamma_X^*(u) = 0.042$ between u_2 and u_3. So u_3 is selected for the post.

Remark 10.17 Cagman et al. [3] gave the fpfs-set decision-making problem by taking $U = \{u_1, u_2, u_3, u_4, u_5, u_6, u_7, u_8\}$ with set of fuzzy parameters $E = \{x_1, x_2, x_3, x_4, x_5\}$ and construct fpfs-set $\Gamma_X = [((0.5)/x_2, \{(0.3)/u_2, (0.4)/u_3, (0.1)/u_4, (0.9)/u_5, (0.7)/u_7\}), ((0.9)/x_3, \{(0.4)/u_1, (0.4)/u_2, (0.9)/u_3, (0.3)/u_4\}), ((0.6)/x_4, \{(0.2)/u_1, (0.5)/u_2, (0.1)/u_5, (0.7)/u_7, (1)/u_8\})]$ where $X = \{(0.5)/x_2, (0.9)/x_3, (0.6)/x_4\}$ of E.

Thus, the aggregate IF set $\Gamma_{X*} = \{(0.096)/u_1, (0.162)/u_2, (0.202)/u_3, (0.064)/u_4, (0.102)/u_5, (0.154)/u_7, (0.12)/u_8\}$.

Finally, the candidate u_3 having the largest membership grade had been chosen for the job.

But we observe that if $X = \{(0.5)/x_2, (0.9)/x_3, (0.6)/x_4\}$ of E and if $\Gamma_X = [((0.5)/x_2, \{(0.3)/u_2, (0.4)/u_3, (0.1)/u_4, (0.9)/u_5, (0.7)/u_7\}), ((0.9)/x_3, \{(0.4)/u_1, (0.5)/u_2, (0.9)/u_3, (0.3)/u_4\}), ((0.6)/x_4, \{(0.2)/u_1, (0.6)/u_2, (0.1)/u_5, (0.7)/u_7, (1)/u_8\})]$, then the aggregate IF set $\Gamma_X^* = \{(0.096)/u_1, (0.202)/u_2, (0.202)/u_3, (0.064)/u_4, (0.102)/u_5, (0.154)/u_7, (0.12)/u_8\}$. Then, u_2 and u_3 have the same largest membership value. So It is not possible to select one between the two candidates. In this case, we need the ifpifs-set theories.

10.2 Conclusion

Cagman et al. [3] gave the fpfs-set decision-making problem. Here, we construct ifpifs-set decision-making problem, which is more fruitful incase when the candidates having the same largest membership-grade $\mu\Gamma_X{}^*(u)$.

References

1. Atanassov, K.: Intuitionistic fuzzy sets. Fuzzy Sets Syst. **20**, 87–96 (1986)
2. Aktas, H., Cagman, N.: Soft sets and soft groups. Inf. Sci. **1**(77), 2726–2735 (2007)
3. Cagman, N., Citak, F., Enginoglu, S.: Fuzzy parameterized soft set theory and its applications. Turk. J. Fuzzy Syst. **1**(1), 21–35 (2010)
4. Feng, F., Li, C., Davvaz, B., Ali, M.I.: Soft sets combined with fuzzy sets and rough sets: a tentative approach. Soft. Comput. Published on line 27th June 2009. doi:10.1007/s00500-009-0465-6
5. Maji, P.K., Biswas, R., Roy, A.R.: Fuzzy soft sets. J. Fuzzy Math. **9**(3), 589–602 (2001)
6. Maji, P.K., Biswas, R., Roy, A.R.: Intuitionistic fuzzy soft sets. J. Fuzzy Math. **9**(3), 677–691 (2001)
7. Maji, P.K., Biswas, R., Roy, A.R.: An application of soft sets in a decition making problem. Comput. Math. Appl. **44**, 1077–1083 (2002)
8. Maji, P.K., Biswas, R., Roy, A.R.: Soft set theory. Comput. Math. Appl. **45**, 555–562 (2003)
9. Majumder, P., Samanta, S.K.: Generalised fuzzy soft sets. Comput. Math. Appl. **59**, 1425–1432 (2010)
10. Molodtsov, D.A.: Soft set theory-first results. Comput. Math Appl. **37**, 19–31 (1999)
11. Roy, A.R., Maji, P.K.: Afuzzy soft set theoretic approach to fuzzy decision making problems. J. Comput. Appl. Math. **203**, 412–418 (2007)
12. Zadeh, L.A.: Fuzzy sets. Inf. Control **8**, 338–353 (1965)

Index

© Springer India 2015
A. Mukherjee, *Generalized Rough Sets*, Studies in Fuzziness and Soft Computing 324, DOI 10.1007/978-81-322-2458-7

Printed in the United States
By Bookmasters